Routledge Revivals

Freshwater Recreational Fishing

The Federal Water Pollution Control Act, signed into law in 1972, dramatically redirected the United States' water pollution control efforts and set out ambitious national goals, expressed both in terms of discharge controls and of resulting water quality. Originally published in 1982, this title examines the benefits that a reduction in the discharge of water pollutants has for recreational fisherman including an increase in the total availability of fishable natural water bodies and an improvement in the aesthetic quality of the fishing experience. It is a valuable resource for students interested in environmental studies and public policy making.

Freshwater Recreational Fishing

The National Benefits of Water Pollution Control

William J. Vaughan and Clifford S. Russell

RFF PRESS
RESOURCES FOR THE FUTURE

First published in 1982
by Resources for the Future, Inc.

This edition first published in 2016 by Routledge
2 Park Square, Milton Park, Abingdon, Oxon, OX14 4RN
and by Routledge
711 Third Avenue, New York, NY 10017

Routledge is an imprint of the Taylor & Francis Group, an informa business

Publisher's Note
The publisher has gone to great lengths to ensure the quality of this reprint but points out that some imperfections in the original copies may be apparent.

Disclaimer
The publisher has made every effort to trace copyright holders and welcomes correspondence from those they have been unable to contact.

A Library of Congress record exists under LC control number: 82016581

ISBN 13: 978-1-138-95438-0 (hbk)
ISBN 13: 978-1-315-66694-5 (ebk)
ISBN 13: 978-1-138-95442-7 (pbk)

Freshwater Recreational Fishing

The National Benefits of Water Pollution Control

WILLIAM J. VAUGHAN and CLIFFORD S. RUSSELL
with the assistance of
CHARLES PAULSEN and RICHARD CARSON

RESOURCES FOR THE FUTURE / WASHINGTON, D.C.

Library of Congress Cataloging in Publication Data

Vaughan, William, J., 1943–
Freshwater recreational fishing.

Includes bibliographical references and index.
1. Fishing—United States. 2. Fishing surveys—United States. 3. Water quality management—United States. 4. Water quality management—Government policy—United States. I. Russell, Clifford S. II. Title.
SH463.V38 1982 333.78'4 82-16581
ISBN 0-8018-2986-0 (Johns Hopkins University Press)

The drawings for the cover and frontispiece were done by Julia C. Allen.

Distributed by The Johns Hopkins University Press, Baltimore, Maryland 21218
Published November 1982

Resources for the Future is a nonprofit organization for research and education in the development, conservation, and use of natural resources, including the quality of the environment. It was established in 1952 with the cooperation of the Ford Foundation. Grants for research are accepted from government and private sources only on the condition that RFF shall be solely responsible for the conduct of the research and free to make its results available to the public. Most of the work of Resources for the Future is carried out by its resident staff; part is supported by grants to universities and other nonprofit organizations. Unless otherwise stated, interpretations and conclusions in RFF publications are those of the authors; the organization takes responsibility for the selection of significant subjects for study, the competence of the researchers, and their freedom of inquiry.

Research Papers are studies and conference reports published by Resources for the Future from the authors' typescripts. The accuracy of the material is the responsibility of the authors and the text is edited at the authors' discretion. Research Papers are intended to provide prompt distribution of research that is likely to have somewhat more limited appeal than that of RFF books.

CONTENTS

List of Tables

List of Figures

ACKNOWLEDGMENTS

We have learned a great deal in the course of this research and would like to thank the many people who helped us to learn it. But first we should express our gratitude to the Environmental Protection Agency, Office of Research and Development, and to Alan Carlin in particular, for the support, in the form of a cooperative agreement between EPA and RFF, which allowed us to undertake the work and the learning.

Many of our colleagues at RFF, who had to listen as our ideas developed, were kind enough to offer their own ideas and answer our questions at crucial times. A complete list of those who took trouble for us would read like the organization's annual report, but we should mention a few by name, especially Leonard Gianessi, Ray Kopp, Charles Paulsen, and Louanne Sawyer.

We were fortunate as well in being able to call on outside consultants for advice in areas of their expertise. Kenneth E. McConnell, University of Maryland, A. Myrick Freeman, III, Bowdoin College, and especially V. Kerry Smith, University of North Carolina, worked to keep us on the econometric straight and narrow. Other useful contributions came from Joseph Yance of EPA and Gardner Brown of the University of Washington. Michael J. Hay of the Fish and Wildlife Service helped us to obtain and use that invaluable data source, the results of the 1975 Survey of Hunting, Fishing, and Wildlife-Associated Recreation. And finally, perhaps the most valuable ideas of all came from Larry Nielsen of the Department of Fisheries and Wildlife Sciences at Virginia Polytechnic Institute and State University. Larry helped direct our thinking about how changes in water quality due to pollution control were likely to show up in fishermen's utility functions and decision processes.

As the reader will discover in the chapters that follow, we asked many factual questions of many people. In particular, we surveyed over 800 private fee fishery managers across the United States and deeply appreciate the time they took to tell us about their operations. We also asked state fish and game officials to give us their estimates of the extent of unfishable water in their states prior to serious implementation of the 1972 water pollution control legislation. We further complicated their lives by asking about the breakdown of fishable water by dominant fish type present. This information provided a crucial baseline for the estimation exercise at the heart of the method report here, and we still marvel at the patience with which these respondents approached the queries of apparently eccentric economists.

Finally, we would like to thank Ruth Haas for editorial assistance and Margaret Parr-Recard and Betty Cawthorne for help in the production of the manuscript.

September 1982

W.J.V.
C.S.R.

Chapter 1

THE PROBLEM AND THE APPROACH

BACKGROUND

One of the most important pieces of national environmental legislation created during the 1970s was the comprehensive amendment of the Federal Water Pollution Control Act (FWPCA(A)). These amendments, signed into law in 1972, in reality constituted a major piece of legislation in their own right, dramatically redirecting the nation's water pollution control efforts and setting out ambitious national goals, expressed both in terms of discharge controls and of resulting water quality. The earlier version of FWPCA (encoded in PL 87-88 of 1961) and the Water Quality Act of 1964 (PL 89-234) set up a system of ambient water quality standards chosen by the states with federal guidance. In contrast to this legislation, the 1972 amendments stressed discharge control (Davies and Davies, 1975). The key elements in this new attack on pollution were a series of general control technology definitions with deadlines for their application.[1]

> Effluent limitations based on best practicable control technology currently available (BPCTCA or BPT for short) were to be applied to all private point sources of waterborne waste discharge by July 1, 1977.
>
> Effluent limitations based on best available technology economically achievable (BATEA or BAT for short) were to be applied to all private point sources of waterborne waste discharge by July 1, 1983.
>
> Effluent limitations based on "secondary treatment" as defined by the administrator (of EPA) were to be applied to publicly owned treatment plants by July 1, 1977. Further, by July 1983, application of "the best practicable waste treatment technology over the life of the works" was to be required.

> The ultimate goal of the national policy was to be the elimination of all pollution discharges to navigable waters by 1985. (This, for obvious reasons, became known as the zero-discharge goal.)

This concentration on requiring discharge controls based on "practicable," "available," and "economically achievable" technologies was meant to avoid problems of translating ambient water quality standards into legally enforceable requirements on particular polluters. But the makers of the 1972 amendments did not entirely abandon direct concern for the ambient environment. In particular, they allowed for interaction between existing state ambient water quality standards and the technology-based treatment standards by requiring more severe discharge reductions than those of BPT if the BPT-based reductions failed to achieve the existing ambient water quality standards (Section 301(b)(1)(C)). In addition, the Congress inserted its own ambient water quality standard in the so-called "fishable-swimmable" goal of Section 101(a)(2): "...it is the national goal that whenever attainable, an interim goal of water quality which provides for the protection and propagation of fish, shellfish, and wildlife and provides for recreation in and on the water be achieved by July 1, 1983." If it will not be met by the application of BAT-based effluent standards, the administrator is to establish stricter effluent limitations that would give the desired result (Section 302).[2]

This piece of legislation covers a great deal more than the definitions and requirements mentioned above. For example, it establishes a mechanism for encouraging regional planning and coordination of pollution control efforts; contains provisions about dredge and fill operations and the protection of wetlands; and defines the design and construction subsidy system for municipal sewage treatment plants--the massive incentive to encourage compliance by unhappy local governments faced with very large costs. However, technology-based treatment requirements and the very aims for ambient water quality, not to mention the zero-discharge goal, have dominated most discussions of the amendments.

Criticisms of the amendments and debate over their goals and requirements began during the legislative process and has continued to the present. The critical positions are that the goals are too ambitious and that the treatment requirements designed to reach the goals are severely flawed, being both highly inefficient in the short run and inimical to technological improvements over the long run. By "too ambitious" the critics mean, and often explicitly say, that the benefits to be expected from meeting the goals (and related requirements) are not large enough to justify the cost of compliance.[3] This line of criticism may be said to have been vindicated, at least to some limited extent, by further amendment of national water pollution control policy in 1977. This legislation relaxed some of the treatment definitions and compliance dates of the 1972 amendments.[4] (It also changed the name of the much-amended law to the Clean Water Act.)

Nevertheless, the debate continues, the relaxations being insufficient to persuade all interested parties that the benefits now clearly outweigh the costs. It is this continuing uncertainty about and disagreement over benefits that may be said to have given rise to the work reported in the following pages, for it is our intention to help inform the debate between critics and defenders of the amendments. In particular, we aim to present both an improved estimate of one part of the full set of benefits attributable to implementation of the legislation, and a method potentially applicable to other parts of the set. In the process we must ignore the other major line of critical argument--the failings of the technology-based standards approach.[5]

The beginning of the decade of the 1980s seems an especially propitious time for this effort because the rhetoric, at least, of the public debate over government action currently stresses the importance of measuring and comparing the benefits and costs of all forms of intervention.[6] In the past, it has appeared that benefit-cost analysis was more a matter of assertion than of analysis. Those who favored the amendments (and other environmental legislation for that matter) often referred to the benefits as "incalculable." Their opponents denied the existence of some categories of benefits and suggested that total benefits, far from being incalculable, were obviously too small to justify the costs implied by the new policies. Neither side had any persuasive and comprehensive estimates to back up their assertions.

SOME PRELIMINARY EVIDENCE: AGGREGATES AND CATEGORIES

We do not wish to exaggerate the current poverty of knowledge of the costs and benefits of water pollution control. There are routine, ongoing efforts to measure costs. While more sporadic, efforts to obtain values for benefits have certainly not been completely lacking. In order to provide some perspective on the importance of this legislation to the economy, a very rough idea of the relative magnitudes of benefits and costs, and guidance about the significance of particular benefit categories, it is worth considering some of the existing evidence. First, however, a few cautionary notes are in order.

While this book is concerned with benefit, not cost, estimation, measuring the costs of the nation's water pollution control program is a far from trivial problem. There are difficulties of definition. For example, there is the question of what costs really are to be attributed to the program and which would have been voluntarily incurred in any case, perhaps to recover a valuable by-product for recycling. On the other side, actual out-of-pocket costs, reflecting public subsidies, must not be confused with real program costs. Beyond definition lies measurement, and here the basic choice is between asking dischargers to report or project their own control costs, and developing "synthetic" costs for hypothetical implementation of the program using engineering-economic estimation of what this would require. Each method has its problems and strengths, but a difficulty facing both is the necessity of having some quite specific idea of how the program, in this case the rather general language of the technology-based standards, will actually be translated into required actions for dischargers.[7]

Bearing in mind that cost estimation and projection is not a simple pro forma exercise, we may examine table 1-1 with a view to understanding both the size of the national commitment represented by FWPCA(A)/CWA, and the distribution of that cost over time and major economic sectors. According to the U.S. Council on Environmental Quality (CEQ), total incremental expenditure on water pollution abatement was $12.7 billion in 1979 (1979 dollars). This is roughly the same amount of spending as that devoted to replacement auto parts and about 90 percent of the sales of the health care and hospital supply industry in 1979.[8] It was about equally divided between public (we may assume primarily municipal) polluters and industrial sources. Electric utilities are estimated to have spent 5.5 percent of the total.

CEQ estimates that by 1988 incremental expenditures on water pollution abatement will have about doubled (in 1979 dollars) to $24.4 billion per year. This represents an average annual rate of growth in real terms of almost 7 percent, considerably higher than even the most optimistic projections of the GNP growth rate. Therefore, this spending will necessarily increase in relative importance, however measured.[9]

Table 1-1. Estimated Incremental Water Pollution Abatement Expenditures 1979-88

(billions of 1979 dollars)

Spending sector	1979			1988			Cumulative 1979-1988		
	Operation and maintenance	Annual capital costs[a]	Total annual costs	Operation and maintenance	Annual capital costs[a]	Total annual costs	Operation and maintenance	Capital costs[a]	Total costs
Public	1.7	4.3	6.0	3.3	10.0	13.3	25.1	59.2	84.3
Private									
Industrial	3.4	2.6	6.0	5.4	4.5	9.9	42.0	34.0	76.0
Electric utilities	.3	.4	.7	.3	.9	1.2	2.9	6.5	9.4
Subtotal private	3.7	3.0	6.7	5.7	5.4	11.1	44.9	40.5	85.4
Total public and private	5.4	7.3	12.7	9.0	15.4	24.4	70.0	99.7	169.7

Note: Table adapted from table 10-1, p. 394, in Environmental Quality: The 11th Annual Report of the Council on Environmental Quality (Washington, D.C., Government Printing Office, 1980). Incremental costs are those made in response to federal legislation beyond those that would have been made in the absence of that legislation.

[a]Interest and depreciation.

If defining and measuring costs is hard, the production of believable benefit estimates for water pollution control is much harder. A first problem is that for the most part these benefits do not involve marketed goods. (Commercial fishing is an exception, but a fairly lonely one.) This is not a matter of happenstance, of course, but a reflection of the fact that for a combination of technical and traditional or customary reasons, the major relevant service outputs of natural waterways (for example, recreation, contribution to scenic vistas) involve public goods--goods available to all, once provided to any. In addition, for another potentially important category of use--drinking water--some major damages of pollution, and hence the benefits of its control, are tangled in the technical difficulties and economic enigmas surrounding human health effects (morbidity and mortality). To make matters even worse, there are persuasive a priori arguments that some important categories of benefits do not involve human <u>use</u> at all, even in the indirect sense of viewing scenery. For example, it has been asserted (and to some limited extent supported by evidence; see Mitchell and Carson, 1981) that people are willing to pay significant amounts just for the knowledge that a pollution control program is resulting in a comprehensive improvement in national water quality. This is quite independent of use, or indeed of intent to use, those waters.[10]

Undaunted if not unhindered by this array of problems, researchers, especially over the past decade, have produced a number of estimates of national water pollution control benefits within several categories. These individual pieces have been brought together, with critical comment and professional judgment, by Freeman (1979a); the results of his efforts are summarized in table 1-2.

A first observation based on tables 1-1 and 1-2 is that on this evidence the benefit-cost balance for water pollution control is a very near thing. Thus, Freeman's "most likely point estimate" for aggregate benefits ($13.9 billion per year in 1979 dollars) is only 9.4 percent higher than CEQ's cost estimate for that year. Even this tentative conclusion must further be tempered with caution. Not only do the ranges reported by Freeman suggest very great uncertainty; his interpretive remarks and examination of individual studies raise another issue: that the "pollution control" assumed for the benefit estimates may not be consistent across benefit categories and is certainly not consistent with the assumptions on which CEQ based its cost projections. Thus, CEQ assumes gradual compliance with the increasingly stringent requirements of the

Table 1-2. National Benefits of Meeting 1985 Water Quality Objectives (billions of 1979 dollars per year)[a]

Category	Range	Most likely point estimate
Recreation		
Fresh water fishing	0.7 - 4.5	1.1
Marine sports fishing	2.3 - 5.7	3.3
Boating	1.4 - 3.2	2.3
Swimming	0.2 - 2.3	0.6
Waterfowl hunting	0.1 - 0.3	0.2
Subtotal[b]	4.6 -16.0	7.6
Nonusable Benefits		
Aesthetics, ecology, property values	1.1 - 5.7	2.3
Diversion Uses		
Drinking water - health	0 - 2.3	1.1
Municipal treatment	0.7 - 1.4	1.0
Households	0.1 - 0.6	0.3
Industrial supplies	0.4 - 0.9	0.7
Subtotal[b]	1.2 - 5.2	3.1
Commercial Fisheries	0.4 - 1.4	0.9
Grand total	7.5 -28.0	13.9

Source: A. Myrick Freeman III, "The Benefits of Air and Water Pollution Control: A Review and Synthesis of Recent Estimates," unpublished report to the Council on Environmental Quality, December 1979; tables 12 and 13; pp. 160, 171, 174.

[a]Freeman's figures were expressed in 1978 dollars and have been converted to 1979 dollars using the Consumer Price Index (CPI). This correction is consistent with that employed by Freeman to put studies from several different years into comparable terms.

[b]Figures may not add to totals and subtotals because of rounding.

legislation, while the figures reported by Freeman essentially all involve one rather complete degree of control and cannot be expanded, partitioned, or extended to cover other possible requirements.[11] In particular, it is probably most defensible to compare adjusted versions of Freeman's benefit estimates with CEQ's annual cost estimate for 1988, by which time, under current law, very stringent treatment will be in place and operating everywhere. This comparison suggests even more strongly that the question of the aggregate benefit-cost balance is far from settled. Thus, granting that benefit estimates based on the same methods as those studies reported by Freeman would show higher benefits in 1988 (still in 1979 dollars) because of higher incomes, more leisure time, and perhaps changing tastes, we would still be very surprised to find projected benefits substantially outweighing projected costs. The near balance of 1979 would surely remain in the adjusted 1988 comparison. In short, it seems clear that a case for or against the current legislation cannot be made decisively with existing information. Because the spending required by that legislation is large, a successful attempt to reduce uncertainty about benefits would be of significant value.

In this connection, it is worthwhile to examine the pattern of benefits by category in Freeman's summary. Of the total annual benefits of \$13.9 billion in 1979 dollars, recreation accounts for well over half, or \$7.6 billion. No other single category even comes close in importance, although the set of categories that appears to cover all aspects of public and private water supply totals \$3.1 billion (22 percent of the total). This suggests that recreational use of water may be a fruitful area to investigate in seeking to reduce the uncertainty of benefit estimates. There are, however, reasons to be puzzled and even skeptical about the subcategory results within "recreation." First, "boating" is supposed to produce 30 percent of the \$7.6 billion subtotal, or \$2.3 billion per year. Given the nature of this activity, and its independence of water quality relative, say, to swimming or fishing, this result strains intuition.

Marine sport fishing looms enormously large in the recreation category, being estimated as three times as important as freshwater angling and accounting for 43 percent of the subtotal. This does not seem to fit with such facts as are available about the relative importance of the two kinds of fishing in terms of activity. In particular, in 1975, fishermen estimated that they spent 20 percent of their total days in saltwater or estuarine fishing. The other 80 percent were spent on freshwater (U.S. Department of Interior, n.d.).

While there is no information on how the values of days spent in various types of fishing might change under the influence of pollution control, Charbonneau and Hay (1978) have estimated the average consumer surplus for fishing days of various sorts. By this measure, the value of an offshore saltwater day is about three and one-half times larger than that for a freshwater day (averaged over species), while the value of a day of saltwater fishing along shore or from shore is roughly the same as that for a freshwater day (Charbonneau and Hay, 1978). Let us say, generously, that each saltwater day is worth three times as much as each freshwater day, and use that relative value in the benefit calculation, ascribing those benefits entirely to increase in the number of days spent in each kind of activity. Then if the increases in days spent in fresh and saltwater fishing were equal, the benefits would stand in the relation shown in table 1-2. But the percentage change in saltwater fishing participation would then be <u>four times</u> as large as that recorded in freshwater fishing. No one could say a priori that this is impossible, but it does provide an additional reason to think that closer examination of the recreational benefits of water pollution control may have high returns in terms of reducing uncertainty.

RECREATIONAL BENEFITS OF WATER POLLUTION CONTROL

In applied research efforts there seems to be a temptation to become enmeshed in the intricacies of traditional theories and techniques, to lose sight of the common sense of the questions involved, and to ignore at the start the implications of both the charge of the project and the capacity of its budget to support intricacy. Here an attempt has been made both to improve methodology for estimating the national recreational benefits of water pollution control and to produce some actual, improved estimates. The budget and time available were insufficient to support a broad attack on recreational benefits generally, and it was decided early in the project to concentrate on freshwater recreational fishing, but within that subcategory to work toward national benefit estimates and the development of a method that would have potentially wider applicability.[12]

The first questions we tried to answer were:

> What are the mechanisms by which control of waste discharges to watercourses creates benefits for actual (or potential) recreational fishermen?
>
> What tools and data would capture those mechanisms?
>
> What special difficulties would arise from an emphasis on obtaining a national benefit total?

The answer to the first question seems to be that a reduction in the discharge of water pollutants has three kinds of beneficial effects for recreational fishermen. It tends to increase the total availability of fishable natural water bodies by reducing the incidence of conditions that make it difficult or impossible for fish to survive, such as very low dissolved oxygen or pH levels, or very high loads of suspended solids. This increases the amount of water available for every actual or potential angler. (Because increased availability may result in more potential fisherman becoming actual fisherman, there may or may not be an overall reduction in average levels of crowding expressed as fishermen per acre or per stream mile.)

The second effect of pollution control comes via changes in the types of fish that can live and reproduce in the affected water bodies. The key observation here (pursued in some detail in chapter 2) is that the fish population shifts that correspond with improvements in dissolved oxygen, suspended solids, and toxic chemical loads, are themselves classified as improvements by anglers. Very simply, clean water means "game" fish (such as trout or bass), dirty water means "rough" fish (such as carp, drum, or buffalo fish), and game fish are generally preferred to rough fish by anglers. Thus, pollution control tends to increase the relative amount of water yielding higher quality fishing.[13]

Finally, even if control of wastewater discharges in a specific water body is not great enough to change its major fish population, there will still be an improvement in the aesthetic quality of the fishing experience. That is, the water will tend, in general, to be clearer and noxious odors and heavy algae growth will be reduced or eliminated.[14]

There may be some losers in this process, because even if the above assertions about angler preferences are true in general there is no reason to expect complete homogeneity. Thus, the avid and diehard carp fisherman may find that pollution control reduces the amount of water available for this pursuit, and the person with a high tolerance for cloudy or smelly water but a low tolerance for crowding may find any improvement in conditions cancelled out by increases in streamside company.[15]

Taking the above as a reasonable view of the benefit-producing mechanisms, how can it be used to help estimate benefits of the nation's water pollution control legislation? One way of answering this question is to catalog five links that we should be able to understand and model (whether formally or informally) in order to predict the benefits of implementation of the law.

> We should be able to predict how implementation will affect pollution discharges by location, quantity, and pollutant type across the entire nation.[16]
>
> We should be able to predict how the pre-policy and post-policy discharge levels affect ambient water quality (or how ambient quality changes as discharges change) in terms not only of such familiar indicators as dissolved oxygen, but in terms of supportable fish population types.
>
> We should be able to predict how increases in total amounts of water supporting recreational fishing and shifts in the composition of that water toward more highly valued fish species affect numbers of anglers and the amount of time they spend fishing.
>
> We should be able to put values on fishing activity of various kinds (that is, for practical purposes, on days spent fishing for various species).
>
> Finally, we should also be able to say how changes in water quality <u>per se</u> affect the value of a fishing day of a given type.

This list suggests some of the formidable difficulties in producing national benefit estimates for even this one subcategory. The first two links alone, from law to discharges and thence to ambient quality, while not necessarily involving subtle economic questions, do require massive amounts of information and predictive mechanisms for water quality in every affected water body in the country.[17] Moreover, as discussed in chapter 2, the prediction of fish populations by type, as a function of physical and chemical water quality parameters, is an underdeveloped art.

Economic considerations enter explicitly when the remaining three links are attacked. Broadly speaking, there are two lines such an attack may take: the places or sites at which recreational fishing is done or the populations that do the fishing. One may think of the site-based method as setting out to measure the net change in the value of all recreational fishing sites across the nation that can be traced to water pollution control policy. (For estimates of annual benefits, the total values of interest would be those of annual service flows.) For a single site, the changes in net value would be given by the change in the sum of producer and consumer surplus due to quality-related shifts in demand and supply functions. This is illustrated in figure 1-1 where the total site surplus expands from abc per year to def per year.

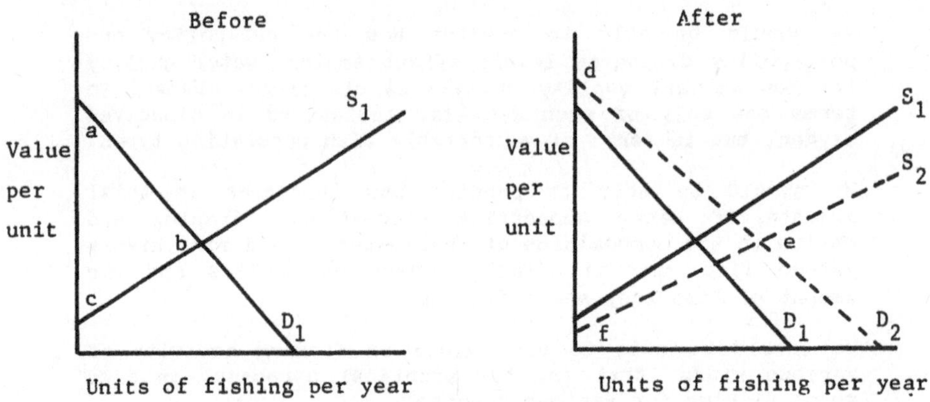

Figure 1-1. Effect on supply of and demand for recreational fishing at a site attributable to water pollution control.

Shifts in demand would occur for the two reasons already discussed--upgrading of available species and improvement in aesthetics. The other factors that might shift the curve are assumed not to change.[18] The necessary demand curves could be estimated by the classical travel-cost demand method, given sufficient time and money for gathering data.

A supply curve may or may not exist for a particular site. A natural site, such as a wilderness pond that has no special provisions for access and no management intervention to increase bag (such as stocking, feeding, or liming), may reasonably be said to provide units of recreational fishing of constant quality at zero marginal cost (bar the problem of congestion just discussed). Management interventions to change any conditions from the natural (or, indeed, to protect the natural by restoration or policing) introduce nonzero marginal costs. Pollution control might shift those costs in either direction. In particular, because of the ambiguity already noted about the sign of the effect of pollution on biomass and hence bag, it is conceivable that a cutback in inflows of organic pollutants might necessitate higher costs to maintain constant bag at any given level of angling pressure.

A particular recreational fishing site exists in a context of other sites, however, and the demand shift assumed above must reflect some explicit assumption about that context. Traditionally, the application of site-specific valuation methods has involved policy questions that are themselves site-specific. (For example, the question typically was: How much demand will there be for a new site to be created by a public works project?) The influence of the other sites making up the context could be taken into account, but the characteristics of those other sites could also be assumed constant between the with (after) and without (before) situations. The evaluation of a comprehensive policy of water pollution control introduces a complication, then, for in general the characteristics of _all_ sites will change as the policy is implemented.[19] This implies that what is involved is a massive general equilibrium problem in which the changes at all sites and the interactions among all the changes must be entered simultaneously in order to determine how the policy shifts demand curves at particular sites.

Another difficulty is logically prior to this formidable problem. The recreational fishing "site" is not a well-defined unit when one looks across all available water bodies in the nation. That is, while it is clear enough that a pond or modest-sized lake constitutes a single site, how are sites on a major river, or a large lake or man-made reservoir defined? While a careful, comprehensive inventory dividing all water bodies into separate sites for which it is reasonable to estimate separate demand curves might be possible, it is far less clear that lacking such a careful study, there is any unit of aggregation corresponding to "site" that allows extrapolation from one, a dozen, or even a hundred specific sites to an estimate of national benefits.

The difficulties mentioned above have suggested to many researchers that a more promising path to the desired goal lies through estimating the effects of pollution control on individual people, each seen as facing a cross section of available sites of varying quality. Here, the object is to find the sum of increases in consumer surpluses, over the individual consumers, that can be attributed to water pollution control via its effect on recreational fishing. A useful way of visualizing this exercise is to begin with household production function theory, where the consumer produces final service flows (or "basic commodities" in the terminology of some authors) using consumption technology, time, and purchased (or market) goods inputs.[20]

In these terms, the consumer has a demand for recreational fishing days (undifferentiated for now) which is satisfied using time, equipment, available sites, and other inputs (Deyak and Smith, 1978). The input availability and consumption technology define the individual's pre-policy supply curve, S_1 in figure 1-2. Implementation of water pollution control reduces the marginal cost to the consumer of producing units of fishing by increasing the availability of the underlying resource, fishable water. Thus, more units of recreational fishing will be "produced" after pollution control. This increase might be estimated using the reduced-form equation for the flow of final services proxied by recreation days, where the exogenous explanatory variables include measures of income, household tastes, prices of particular related market goods, and some measure of resource availability.

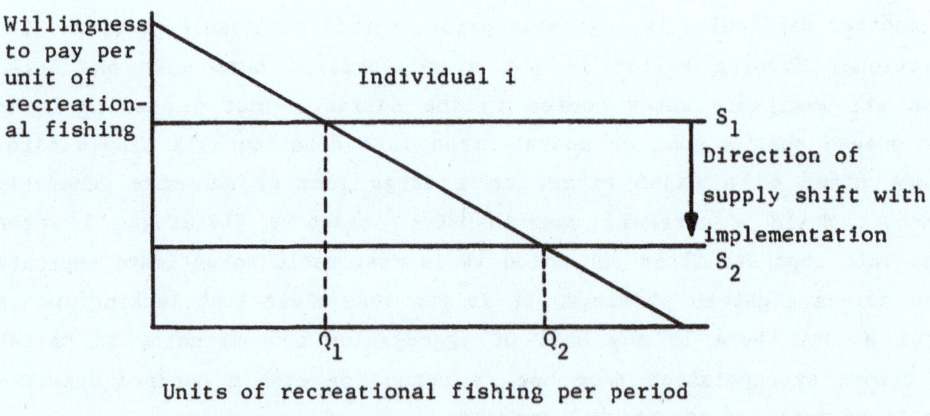

Figure 1-2. Individual's changing production of days of recreational fishing with shift in resource availability.

The situation is somewhat more complicated when the fishing days are to be distinguished on the basis of the quarry sought. If we continue to view changes in resource availability as exogenous to individuals, however, we can continue to think of the reduced-form equations for fishing days (final service flows), one equation for each type of day, in which the entire menu of resource availabilities is represented in each equation. The first effect, then, of water pollution control is, on net, an increase in the amount of recreational fishing done. In addition, there is likely to be a shift in type of fishing or a net substitution of game for rough fishing.

In fact, in this study the reduced-form equations for fishing days are represented by a three-stage process reflecting both the available data and the hypothesized nature of the decision process. That is, we divide the overall "production of days" decision into three parts:

(1) whether or not to fish in a particular year at all;

(2) given an affirmative decision in (1), a decision whether to do some cold-water game fishing, some warm-water game fishing, or some rough fishing (or some combination of them)

(3) how many days to devote to each.

Each part of the decision is hypothesized to reflect the availability of fishery resources, whether as total fishable acres per capita (1); shares of fishable water by fishery type (2); or acres of fishable water per capita, by type of fishery (3).

In order to translate increased activity into benefits, we have to know what each unit of activity is worth, and this raises questions and difficulties for the underlying model. We are visualizing each individual as adding recreational units at his or her "internal" margin, because of the downward shift of the supply function. The appropriate value for the added units is then the marginal unit value, unless so many are added that the change can no longer reasonably be taken to be "marginal." However, the marginal value for each individual is not usually observable.

One possibility for getting at this value is to ask well-constructed questions about willingness to pay (or to accept compensation) for a one-day change in the annual number of fishing days by type.[21] An average of the responses might be used in practical valuation. Another possibility would be to look for evidence of what is actually paid for a representative experience of a particular fishing type. Such evidence is not commonly generated because most recreational fishing takes place at publicly provided sites where the fishing experience is provided free or for a nominal fee.

As a practical matter, the data available in studies of recreational activities will almost never allow one to determine where each individual falls along the spectra of initial (pre-policy) and final (post-policy) equilibria--from severely constrained availability and high marginal values to abundance of opportunities and low marginal values. The imputation of an average willingness-to-pay value to the sum of all individuals' changes in participation, then, is an intuitively appealing approximation. Appendix A develops this argument in more formal terms and provides a demonstration of the accuracy of the approximation under different assumptions about functional form, size of availability change, and pre-policy consumption level. To summarize these latter results:

> If the representative consumer/producer of recreation days has a linear demand function, then the average pre-policy consumer's surplus valuation rule will understate the benefit of the policy change, with the understatement worse the larger the change in activity days. For small changes in quantity, the approximation is very close to one half the true value of the consumer surplus addition.[22]
>
> If the representative consumer's demand is characterized by constant price elasticity, then the approximation is better the smaller the policy-related change in availability (with that change positive under most circumstances).[23] The approximation overstates benefits for low price elasticities, is very close for elasticities around 1.5, and understates benefits for higher price elasticities. For an elasticity of one, and a small change in quantity demanded (that is, where the ratio of post-policy to pre-policy days is within 5 percent of one) the overstatement of the approximation is about 30 percent.

Granting, then, that the use of an estimate of average, pre-policy consumer's surplus may produce quite good approximations to actual benefits of a water quality change, we still must find those original estimates. As mentioned above in the individual context, we could use willingness-to-pay surveys or we could tease the answer out of the data generated by for-fee recreational operations.[24] A third possibility is to adopt the travel cost demand curve, already suggested under the site-based method. Here the objective would not be a site value but an experience value at a representative site offering a particular type of fishing. To obtain this, the estimate of consumer's surplus for such a site could be averaged over the total units of recreation involved.

Implicit in the above method is the argument that because individuals, both before and after policy implementation, face a range of site availabilities with differing water qualities and fishing types, the comprehensive shift in water quality after implementation can be handled as merely a change in quantity rather than a change for each individual in a distance- and direction-specific site menu. This allows us to avoid the general equilibrium adjustment problem. In addition, aggregation is over individual people, presenting no special problems of definition or measurement.

The source of benefits untouched by this general line of attack is the increasing value of a fishing day of particular type (game or rough) because of improving water quality *per se*. (That is, where water quality enters the individual's utility function directly.) In principle, the valuation method outlined above could be expanded to reflect increasing average consumer's surplus due to higher water quality independent of fishing type. In practice, the necessary comprehensive data on sites and their characteristics, including water quality parameters and visitors, could not be obtained within this project's budget.

One further general observation is pertinent. The technique set out here can be applied to other subcategories of water-based recreation, provided one has or can obtain data on:

> the baseline and projected levels of water quality relevant to subcategories across the nation, by geographic or political subdivisions.
>
> participation rates for the activity in question that can be connected to the data on the availability of water of different quality levels, usually via the participants' place of residence.
>
> values per unit of participation, or sufficient data from which to estimate such values, as from a travel cost model. (If this data set includes site water quality differences with enough variation, and is sufficiently large and rich in other dimensions, it may be possible to show how value changes with water quality.)

APPLICATION TO RECREATIONAL FISHING

The chapters that follow describe in detail the application of this method to recreational fishing. The structure of the application is sketched in figure 1-3, which shows that the method requires at a minimum the following data, estimation, and modeling components:

Data on:

the availability of water quality in the base (pre-FWPCA(A)/CWA) period. This involves an inventory of the nation's freshwater reflecting both whether the water body is fishable or not; and if fishable, what kind of fish population dominates. The geographic fineness of this inventory should at least match the fineness of knowledge about individual decisions. (That is, it should at least be possible to connect individuals with water quality availability by political subdivisions, such as the state.)

the decisions of individuals facing this availability. This involves, for recreational fishing, information on the socioeconomic characteristics of a national sample of individuals and information about the choices made by each individual to fish or not; to fish for rough or game fish; and how many days to spend fishing for cold-water and warm-water game fish and rough fish.

the net willingness-to-pay valuations of fishing days of the relevant types.

Estimation of:

prediction equations connecting water availability, by fishery type, with individual decisions, including the decisions whether or not to fish, what kind of fishing to do, and how many days per year to devote to the several possible types of fishing.

The stages of the individual's decisions follow the framework suggested by Cicchetti and coauthors (1972). In all stages the individual's decisions are hypothesized to be sensitive to the availability of water quality suitable for the activity.

Stage I. Estimation of an equation for predicting probability of being a fisherman

Stage II. Estimation of equations for predicting the probability of doing some:

a. Cold-water game fishing (trout)
b. Warm-water game fishing (bass or pan fishing)

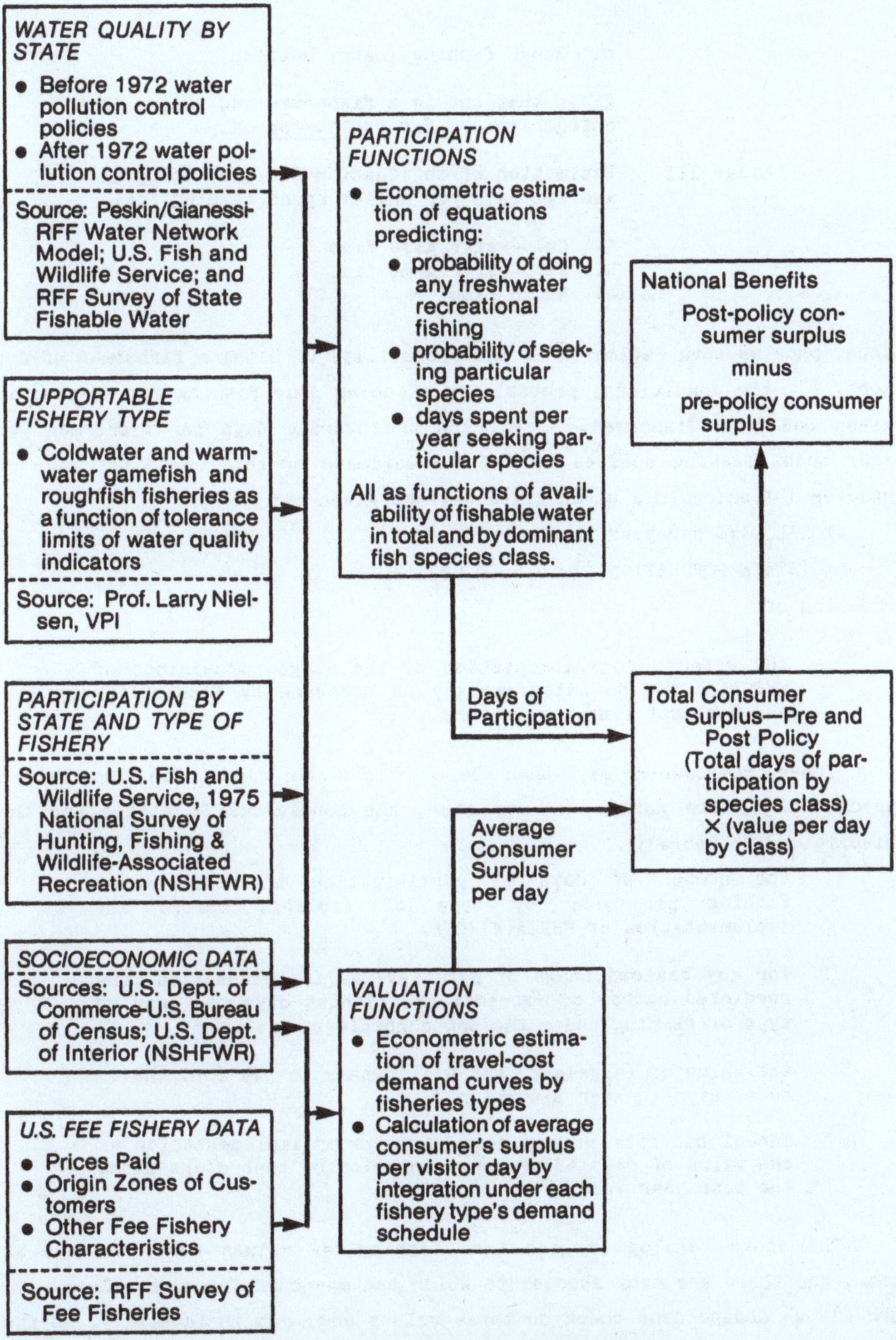

Figure 1-3. Estimating national recreational fishing benefits from water pollution control: A schematic view

c. Rough fishing (carp, catfish)

given that one is a fisherman and noting that categories are for freshwater only.

Stage III. Estimation of an equation for predicting the average days per angler spent fishing for:

a. Cold-water game fish
b. Warm-water game fish
c. Rough fish

Thus, once we have estimates of the probability of being a fisherman of any sort, P_f, the conditional probability of doing some fishing for species i, given one is a fisherman, $P_i | P_f$, and the average days per fisherman per year spent seeking species i, D_i, the estimate of total days per year of species i fishing in a given state will be given by:

TOTAL DAYS per year in state j

= (STATE POPULATION j) (P_f) $(P_i | P_f)(D_i)$

Modeling of:

the effect of implementation of the staged provisions of FWPCA(A)/CWA on water quality as measured by fishability and dominant fish populations.

These components may then be combined to find, for the entire population of the nation (in our case, the contiguous 48 states and the District of Columbia):

the number of days of participation in recreational fishing per year by type of fishing, before any implementation of FWPCA(A)/CWA.

for any desired stage of FWPCA(A)/CWA implementation, the predicted number of recreational fishing days per year by type of fishing under the new conditions.

the value of (aggregate net willingness to pay for) the total days in each situation.

annual benefits of any desired stage of implementation as the value of days after implementation of that stage minus the base year value.

The above catalog assumed that fishing-day values were obtained as data, and there are many studies to which one may turn for such values. To provide an independent check on these values and, not incidentally, on the basic assumption about the scale of angler preferences over fishing types, we made our own estimates of net willingness to pay for days of fishing of various types. To do so, we gathered data on the characteristics, especially fishery type, of several hundred recreational fishing sites across the country. These sites charged a fee, so that some data on numbers of users and on the origins of users by distance from the site were "pre-collected." We then used these data in a one-step analog of a two-stage travel-cost demand-curve estimation process (Freeman, 1979b, pp. 212 et seq.; Granger and coauthors, 1979) of the site-specific area model type as classified by Cicchetti, Fisher, and Smith (1973). That is, in simple terms, we first used our data on visitor frequency at each site, zones of visitor origin by distance from each site, and average characteristics of the populations of each of the zones around each of the sites, to estimate a visitation equation of the form for each site:

$$V = a_i + b_i TC + c_i E$$

where V is visitors per year to the ith site from a travel-cost zone; TC is travel cost from the center distance of the zone to the site; and vE is the vector of characteristics of the population of the zone. Then, in the second step, the characteristics of each site were used to explain the variation in the parameters of the visitation equation across sites; thus:

$$\begin{aligned} a_i &= d + eS_i \\ b_i &= f + gS_i \\ c_i &= h + kS_i \end{aligned}$$

where S_i is the vector of site i's characteristics.

Our actual estimation procedure combines the two stages into one to produce an equation for estimation in which the visitation rate at a site is a function of site characteristics, travel costs, zone population characteristics, and the interaction terms among site characteristics and the independent variables of the simple visitation equation:

$$V = d + eS + fTC + hE + gS(TC) + kS(E)$$

This equation may then be manipulated to produce the demand function for a site of particular fishery type, representative other characteristics, with population means for the characteristics of the travel cost zones. From this function, in turn, consumer's surplus and average consumer's surplus per visitor day may be calculated.

PLAN OF THE BOOK

The study components briefly set out above are described more carefully in the chapters that follow. In chapter 2 we discuss both the sources of our baseline water quality availability data and our method for projecting the changes in this key variable that result from assumed implementation of particular stages of FWPCA(A)/CWA. Chapter 3 deals with the data and estimation techniques required to produce the basic prediction equation for the individual's fish/do-not-fish decision. The prediction equations for particular types of fishing and for the number of days spent per year in those types are dealt with in chapter 4. The valuation of those days is taken up in chapter 5. The components are combined in chapter 6, where we calculate benefits of the implementation of several different stages of FWPCA(A)/CWA under different assumptions about certain key parameters. Chapter 6 also discusses the implications of our results and methods both for the debate over water pollution control policy and for further attempts to improve benefit estimates.

NOTES

1. The full text of the 1972 amendments, with subsequent reamendments from 1977 and 1978, is found in Government Institutes, 1980.

2. Interestingly enough, Section 302 calls for benefit-cost analysis. The language requires such analysis of the stricter effluent guidelines, but implicitly it must concentrate on the incremental improvement in ambient water quality to "fishable-swimmable" from whatever would result from application of BAT. If there were a demonstration that there is "no reasonable relationship between costs and benefits obtained," the stricter effluent limitations would not apply (Section 302(b)(2)).

3. See, for example, American Iron and Steel Institute (1980, pp. 68-71).

4. For example, Section 301(b)(2)(A) was amended to remove the July 1, 1983 deadline for BAT. The BAT treatment definition would still apply to toxic pollutant discharges, but conventional pollutant discharges were made subject to a "best conventional pollutant control technology" as defined by the administrator of EPA, to be effective by July 1, 1984. This was clearly meant to represent a relaxation of both timing and stringency of the treatment standard.

5. For valuable discussions of technology-based standards generally, and as applied in this country specifically, see Freeman (1980). For further discussion of FWPCA(A) and its problems, see Freeman (1978).

6. Most pertinently, see President Reagan's Executive Order 12291 of February 17, 1981, requiring that no major federal agency regulatory actions, "within the discretion allowed by law," should be taken unless potential benefits to society outweigh potential costs; and that regulatory priorities be set on the basis of net benefits for society (A/WPR, 1981).

7. For more complete discussions of the costs of pollution-control programs, and the problems in their measurement, see Gianessi and Peskin (1980) and Portney (1981).

8. Because of the definitions used to construct the national economic accounts, it would be misleading to say that this represents 0.6 percent of 1979 GNP, for not all this spending would appear in the GNP account (see Peskin, 1981). But $12.7 billion is about 0.6 percent of $2,300 billion, roughly the 1979 GNP.

9. If CEQ is proved correct on costs, and if GNP grows at 2.5 percent on average in real terms, incremental spending on water pollution abatement will be 0.8 percent as large as GNP in 1988.

10. In practice, it may be extremely difficult to disentangle this "knowledge" or "existence" benefit from the "option-value" benefits that result from the assurance that future options to use water bodies

recreationally or scenically have been opened up, or at least not foreclosed. A further complication is ecosystem preservation, which some assert is a matter of future human survival and others see in terms of future choices of recreation or scenery, and future scientific knowledge.

11. There is an additional problem here as well. Projections of future costs and benefits necessarily involve assumptions about such difficult matters as industrial growth, per capita consumption, technical change, and the other components of the overall *ceteris paribus* condition.

12. While we have argued in the text that it is hard to accept the benefit estimate of marine recreational fishing, nonetheless, it would clearly have been desirable at least to add, if not concentrate on, this category of benefits. However, we were unable to discover the data and models necessary to link in any convincing way implementation of the FWPCA(A)/CWA to changes in water quality (let alone water quality in terms meaningful to recreational fishing) in estuarine and close-inshore marine waters. Nor, indeed, could we find estimates of existing waters in those areas too polluted to allow fishing, or where fishing quality could be said to be obviously subnormal because of pollution.

    Bell and Canterbery (1976), whose report is the basis for Freeman's marine recreational fishing benefit estimates, used baseline water quality inventory information from the STORET system. They had no quantitative connection between these data and fish types or availabilities. They assumed linear improvement in ambient water quality, approaching their version of perfection, as FWPCA(A) was implemented. They further assumed that this improvement gave rise to a nonlinear increase in productivity of fish populations.

13. As discussed in chapter 2, we assume that if such population shifts can take place, they will take place, speeded along by state fish and game managers who share and reflect the commonly accepted view of angler preferences.

14. There is ambiguity about the potential effect of reduction in pollution discharges on bag (catch per fishing day per person), at least over small ranges. To the extent that pollution consists of organic material and nutrients, the aquatic ecological system can convert it into increased biomass in every system "compartment," including the fishes. Thus, *ceteris paribus*, there may be more fish available in slightly polluted water than in water receiving no discharges. Of course, toxic materials and suspended solids would work in the other direction, and excessive organic discharges would result in oxygen depletion and perhaps other stresses.

15. There may still be carp or other rough fish surviving alongside new (or newly dominant) game fish populations, but there are likely to be fewer than before the change, and one would expect them to be harder to catch, given that game fishermen are now flogging the same water. Our method is based on changes in dominant fish species, however, and we do not attempt to model these complex and subtle interactions.

16. Ideally, we should also know the timing of discharges and any changes in timing attributable to the law.

17. Notice that even the apparently easy way out--seizing on the "fishable/swimmable" goal and beginning with this as the result of implementation--does not entirely remove the difficulty. Even leaving aside the fact that this approach can at best produce a single point estimate and thus cannot inform the debate about what the appropriate level of control is, the method still requires comprehensive knowledge of conditions prior to implementation. Thus, it is not necessary to do nationwide water quality modeling, but a full inventory of base-year water quality is needed. Note also that "fishable," being nonspecific about fishery type and background water quality, implicitly leaves out of consideration benefits we hypothesize result from upgrading of species and higher water quality per se.

18. One of these factors introduces some difficult problems in principle. Congestion (as in users per bank mile) is usually deemed a priori to be an important site characteristic. But if demand increases, at each unit value there will be more units of use per period and hence necessarily more congestion. Thus, if the site demand curve is defined for a given level of congestion, only one point on it can be consistent with that assumed level. All other points imply more or fewer users per period than the assumed congestion level. (See Freeman and Haveman, 1977, for a discussion of this problem in a slightly different context.)

19. As a practical matter, significant change in water quality from the implementation of FWPCA(A)/CWA will be far from universal. The Council on Environmental Quality, in 1971 (CEQ, 1971, p. 220), estimated that over 75 percent of all stream miles in the United States were "unpolluted" in 1970. (It is unclear whether this figure is for 50 states or the contiguous 48 states.)

20. An alternative theoretical justification is provided by Cicchetti, and others (1972), especially pp. 845-848.

21. This is roughly the method represented by the 1975 Fish and Wildlife Service, U.S. Department of the Interior survey (n.d.), where the instrument asked:

> Having thought about how much this activity cost you in 1975, how much more money would you spend annually on your very favorite activity before deciding to stop doing it because it is too expensive?"

The willingness-to-pay valuation of a day's sport, by type of fishing or hunting involved, was calculated from the answers to this question. See, for example, Charbonneau and Hay (1978) and chapter 5 below. Unfortunately, the willingness-to-pay responses reflect the area under each individual's demand curve for fishing service flows, not his marginal cost of producing them. Thus, we have no appropriate measure of the dependent variable to use in the reduced-form equation for the marginal unit value (equal to marginal cost) of the service flow in Deyak and Smith's unit value reduced-form equation (1978, p. 69). Even if it were possible somehow to produce marginal unit values from this response (perhaps via the techniques discussed in Charbonneau and Hay,

1978) only a subset of all participants in recreational fishing provided value information, and those who did provided it only for the type of fishing that was their very favorite activity of all the possibilities on the survey (which included several types of hunting as well as saltwater fishing).

22. The change in activity days results from the change in availability, and therefore this statement might equally well be written in terms of the change in availability. We often do so in what follows.

23. The constant elasticity form for a system of demand functions for all commodities in the household's budget is inconsistent with the adding-up condition of demand theory. However, for the analysis of a single commodity, the constant elasticity form is not necessarily inappropriate because we are not imposing this functional form on the demand functions that remain unspecified (Deaton and Muellbauer, 1980, p. 17).

24. We began this study with this intention in mind, but found it impossible to obtain data of sufficient quality in sufficient amounts to estimate the required demand functions. See chapter 5.

REFERENCES

Air/Water Pollution Report (A/WPR). 1981. "Reagan Orders Federal Regulators to Choose Least Burdensome Course," (February 23) p. 73.

American Iron and Steel Institute. 1980. Steel at the Crossroads (Washington, D.C., AISI) 92 pp.

Bell, Frederick W. and E. Ray Canterbery. 1976. Benefits from Water Pollution Abatement. A report to the National Commission on Water Quality (Tallahassee, Florida State University). Available from NTIS as PB 252 172-01; 217 pp.

Charbonneau, John and Michael J. Hay. 1978. "Determinants and Economic Values of Hunting and Fishing." A paper presented at the 43rd North American Wildlife and Natural Resources Conference, Phoenix, Arizona, March 18-22, 20 pp.

Cicchetti, Charles J., V. Kerry Smith, Jack L. Knetsch, and R. A. Patton. 1972. "Recreation Benefit Estimation and Forecasting: Implications of the Identification Problem," Water Resources Research vol. 8, no. 4 (August) pp. 840-850.

______, Anthony C. Fisher, and V. Kerry Smith. 1973. "Economic Models and Planning Outdoor Recreation," Operations Research vol. 21, no. 5 (September-October) pp. 1104-1113.

Council on Environmental Quality. 1971. Environmental Quality (Washington, D.C., CEQ) 360 pp.

Davies, J. Clarence III and Barbara Davies. 1975. The Politics of Pollution (2nd ed., Indianapolis, Ind., Pegasus) pp. 27-44, esp. p. 30.

Deaton, Angus and John Muellbauer. 1980. Economics and Consumer Behavior (New York, Cambridge University Press).

Deyak, Timothy A. and V. Kerry Smith. 1978. "Congestion and Participation in Outdoor Recreation: A Household Production Function Approach," Journal of Environmental Economics and Management vol. 5, no. 1 (March) pp. 63-80.

Freeman, A. Myrick III. 1978. "Air and Water Pollution Policy," in Paul R. Portney, ed., Current Issues in U.S. Environmental Policy (Baltimore, Md., Johns Hopkins University Press for Resources for the Future) 207 pp.

______. 1979a. "The Benefits of Air and Water Pollution Control: A Review and Synthesis of Recent Estimates," unpublished report to the Council on Environmental Quality (December).

______. 1979b. The Benefits of Environmental Improvement (Baltimore, Md., Johns Hopkins University Press for Resources for the Future).

______. 1980. "Technology-Based Effluent Standards: The U.S. Case," Water Resources Research vol. 16, no. 1 (February) pp. 21-27.

______ and Robert H. Haveman. 1977. "Congestion, Quality Deterioration and Heterogeneous Tastes," Journal of Public Economics vol. 8, no. 2 (October) pp. 225-232.

Gianessi, Leonard P. and Henry M. Peskin. 1980. "The Distribution of the Costs of Federal Water Pollution Control Policy," Land Economics vol. 56, no. 1 (February) pp. 85-102.

Government Institutes. 1980. Environmental Statutes: 1980 Edition (Washington, D.C., Government Institutes) March, pp. 185-310.

Granger, Clive W. J., Robert Engle, Ramu Ramanathan and Alan Anderson. 1979. "Residential Load Curves and Time-of-Day Pricing," Journal of Econometrics vol. 9, pp. 13-32.

Mitchell, Robert Cameron and Richard Carson. 1981. "An Experiment in Measuring Willingness to Pay for Intrinsic Water Pollution Control Benefits." A report to the U.S. Environmental Protection Agency (Washington, D.C., Resources for the Future).

Peskin, Henry M. 1981. "National Income Accounts and the Environment," in Henry M. Peskin, Paul R. Portney and Allen V. Kneese, eds., Environmental Regulation and the U.S. Economy (Baltimore, Md., Johns Hopkins University Press for Resources for the Future).

Portney, Paul R. 1981. "The Macroeconomic Impacts of Federal Environmental Regulation," in Henry M. Peskin, Paul R. Portney and Allen V. Kneese, eds., Environmental Regulation and the U.S. Economy (Baltimore, Md., Johns Hopkins University Press for Resources for the Future).

U.S. Department of the Interior. n.d. 1975 National Survey of Hunting, Fishing, and Wildlife-Associated Recreation (NSHFWR) Data tapes and unpublished mimeo, state and national reports (Washington, D.C.).

CHAPTER 2

MEASURING AND PREDICTING WATER QUALITY IN RECREATION-RELATED TERMS

In order to estimate the monetary benefits accruing to recreational water users from nationwide implementation of water pollution control policies, five major links must be captured quantitatively: (1) those translating the policies into levels of pollutant discharges by sources; (2) those translating the discharges into ambient environmental conditions via natural mechanisms for transportation, dilution, and transformation of pollutants; (3) those translating ambient changes in scientific terms, such as dissolved oxygen, into terms readily perceived and acted upon by recreators; (4) those governing the patterns of responses by recreators (in terms both by intensity and type of participation) to changes in perceived ambient environmental quality; and (5) those involved in the valuation of the recreators' responses. This chapter focuses on the practical aspects of quantitatively representing the essence of the first three of the above links, which are logically prior to the participation and valuation steps.

The benefits of water pollution abatement associated with recreational fishing activity involve individual choices in a central way, because these benefits accrue only to those individuals who choose to participate in the activity and decide how frequently to participate over the year. Each of the estimated participation equations discussed in chapters 3 and 4 is associated with one of the steps in the individual's decision process: whether or not to fish at all, what kinds of fish to seek, and how many days per year to spend seeking them. Fundamental to each of these three equations is the hypothesis that such choices are sensitive to the <u>availability</u> of the opportunity to fish at all and to fish for specific species groups. The availabilities are represented as explanatory variables in the participation equations by either the total acres of fishable water per capita (by state) or the division of that water area into three classes, depending on the broad class of fish populations that it is capable of supporting: cold-water game fish; warm-water game fish-panfish; and warm-water and cold-water food fish-forage fish (to which we refer loosely in the following discussion as rough fish).

Therefore it is necessary to have baseline data on availability of fishable water (and division into fishery classes) in order to estimate the participation equations. Further, in order to measure the benefits of the nation's water pollution control program, we must have a way of predicting how implementation of that program will change both total water availability and its division into classes.

As a practical matter, it is difficult to translate national legislation mandating a system of technology-based pollution discharge standards, expressed as state-issued permits for individual polluters (principally industrial establishments and municipal sewage treatment plants), into predicted changes in water quality conditions of the particular sort needed here. The difficulties are compounded by the presence in waterways of man-induced, nonpoint source pollution such as eroded farm soil and accompanying fertilizer and pesticide residues. In principle, however, the logic of the translation runs as follows:

(1) The legislation is translated into federal (EPA) regulations.

(2) The regulations become the basis for state permitting programs.

(3) Permits are issued to sources requiring them to meet specific discharge limits.

(4) The resulting reduction in discharges causes improvements in ambient water quality as measured by such parameters as dissolved oxygen (DO), total suspended solids (TSS) and pH.

(5) The changes in physical and chemical measures of water quality make possible changes in aquatic ecosystems. In particular, the types of fish populations that can survive and reproduce in the affected bodies or stretches of water will change.

This chapter has two purposes. The first is to introduce data on the pre-policy availability of fishable water and its distribution across supportable species classes on a state-by-state basis. These data will be used in estimating the participation equations discussed in the next two chapters. The second purpose is to describe a quantitative way of capturing the links between the policy and its ambient effects measured in recreator-specific terms. This mechanism is a water quality simulation model with both source discharge and natural systems components, designed to predict the ambient effects of increasingly stringent water pollution control policies above a 1972 pre FWPCA(A)/CWA policy baseline. Model forecasts of the availability of fishable water by species class in the United States after the application of increasingly stringent control policies are provided on a nationwide basis at the end of the chapter.[1] These forecasts are used to produce benefit estimates in chapter 6.

SURVEY DATA DESCRIBING BASELINE RECREATIONAL FISHING CONDITIONS

We were fortunate to find a published, albeit slightly dated, source of information on total acreage of fishable water by state in the United States. This we refer to as Hatchery Fish (U.S. Department of the Interior, 1968). It is a document based on survey data collected in 1966 from federal and state conservation agencies on the amount and types of sport fish habitat; amount and types of habitat stocked; numbers and types of fishermen; stocking requirements; and hatchery production capabilities.

This data source is not especially helpful on the allocation of fishable water to fishery classes, confining itself to a division into cold and warm water, with no finer division of either of those two broad categories. The information on total availability is, however, exactly what is needed for the estimation of the fish/do not fish equation discussed in chapter 3, and, in addition, provides a useful benchmark for making judgments about and adjustments to the results of our allocation and forecasting techniques.

In order to obtain both a baseline allocation to fishery classes of fishable water in each state, and a check on our forecasts of this allocation after pollution-control, we conducted an informal survey of people in state fish and game and environmental protection agencies, supplemented by the opinions of experts at a number of universities. The questionnaire asked for information on the base period (pre-FWPCA(A)/CWA) availability in the official's state of fresh water for recreational fishing, excluding the Great Lakes. We asked about the percentage allocation of total fresh water between the fishable and nonfishable categories and also about the percentage distribution of the fishable water among five dominant types of fish populations which we collapsed in our applied work into three categories:[2]

Survey Category	Aggregated Category
1. Cold-water game fish	1. Cold-water game fish
2. Warm-water game fish 3. Warm-water panfish	2. Warm-water game fish and panfish
4. Food fish (e.g., catfish) 5. Rough fish	3. Food fish (catfish) and rough fish

We further asked for the respondents' predictions of the effect of implementation of FWPCA(A)/CWA.

The response rate to this survey was very high. Only six states out of forty-eight (or 12 percent) failed to provide us with at least one completed questionnaire. The combined data from Hatchery Fish and our own Survey of Fishable Water represent the raw material for estimating the participation equations. Unfortunately the accuracy of the responses is suspect, since the quality of the completed forms appeared to vary considerably across states, and the provisional character of the estimates was emphasized by many of the respondents.[3] This uncertainty raises a potential problem of errors in measurement of the key explanatory variables in the second and third participation equations, which means that the parameter estimates for these two equations may be biased and inconsistent (Pindyck and Rubinfeld, 1976). While we offer no econometric "solution" to this problem, we do perform an informal sensitivity analysis. In chapter 4 we compare the parameter estimates for the water availability variable obtained from the equations using the baseline policy variable values obtained from our Survey of Fishable Water to two alternative sets of "synthetic" or "model generated" values based on the RFF Water Quality Network Model using the Nielsen allocation rules (both described below) whose principal use in this study is for forecasting the values of the key policy variables as a result of implementation of FWPCA(A)/CWA.

Before presenting the data on baseline allocations we discuss the "synthetic" method of generating similar baseline data, which also produces post-policy forecasts--the water quality simulation model.

SYNTHETIC DATA DESCRIBING BASELINE AND POST-POLICY RECREATIONAL FISHING

The RFF Water Quality Network Model, WQNM, discussed in detail in Gianessi, Peskin and Young (1981) and Gianessi and Peskin (1981) was used to translate pollution control legislation into ambient effects.

The model consists of the following elements:[4]

A very large inventory of point sources of water pollution (over 3,000 industrial plants and over 24,000 municipal sewage treatment plants)

An inventory of nonpoint water pollution loads by county

A water quality network model in which the pollution loads are translated by simplified biochemical and physical models into measures of ambient water quality for segments of major rivers and associated reservoirs throughout the U.S. There are 1,051 river points in the model representing 44,507 river miles, along with 175 lakes

A systematic set of rules for applying technology-based discharge standards to the point sources and "best management practices" to the nonpoint sources to mimic operation of the legislation.

Our modification of the model was to add a final step which translates the model's physical measures of ambient environmental quality (dissolved oxygen, etc.) into recreator-perceived measures defined in terms of the fishability of the water, classified by the dominant species class which can be supported (Nielsen, 1980).

At this point a sketch of the structure of the overall model might help to introduce terminology efficiently and assist the reader in visualizing the interconnections. Such a sketch is provided in figure 2-1, whence it can be seen that, for any stream covered by the WQNM, point source discharges are inserted and ambient quality measures are predicted at <u>nodes</u>. The stream lengths between nodes are the <u>reaches</u>. Nonpoint pollution loads are introduced uniformly along the reaches. The decay of oxygen demanding organic material and the consequent consumption of in-stream oxygen are modeled for each reach.

Each of the model's elements is discussed in greater detail below.

<u>Baseline Discharge Inventory</u>

Estimates of pollutant discharges from point source sectors were made for 1972 levels of control. The basic approach for most point source sectors was to assemble a plant inventory showing county location and plant output

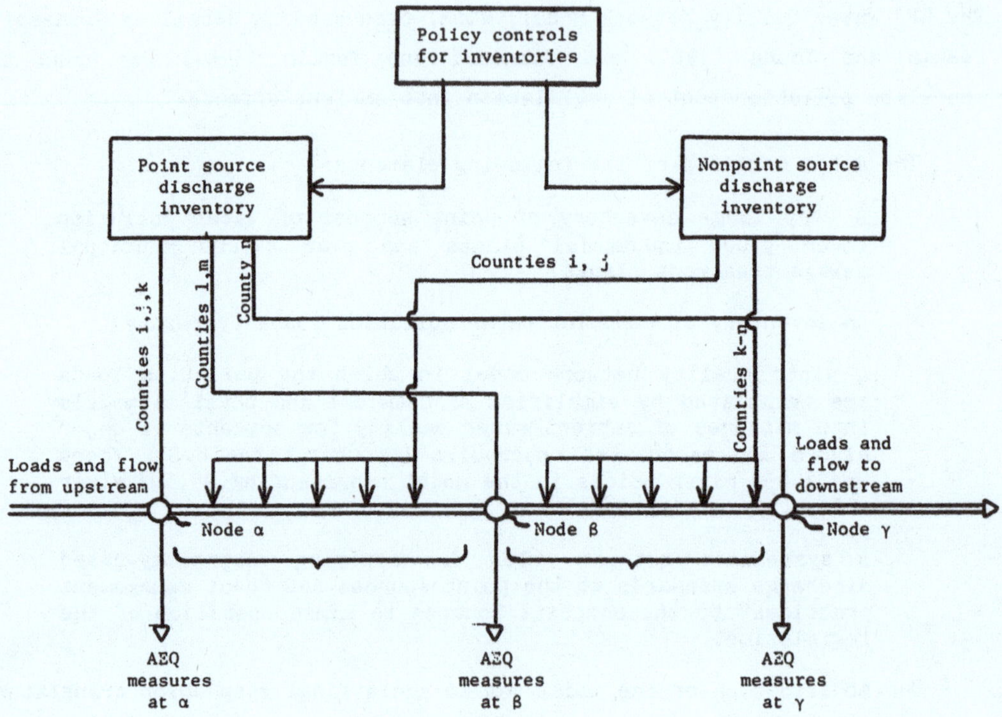

Figure 2-1. RFF water quality network model: Schematic structure for one stream

for all plants, by industry, in 1972. Waste coefficients, based on survey data from actual plants developed by the Environmental Protection Agency, were then used to estimate total waste generated at each plant in 1972, assuming no end-of-pipe pollution control. The actual end-of-pipe wastewater treatment equipment in place in 1972 was then determined for each plant wherever possible. Standard removal efficiencies associated with this equipment were assumed and application of these efficiencies to estimated raw loads produced estimates of actual discharges for 1972. The basic source of data used in this project was a set of technical reports known as Development Documents, prepared by the Effluent Guidelines Division of the Environmental Protection Agency (U.S. EPA, 1974).

For sectors where complete plant inventories could not be assembled, a cruder method of assigning sector-by-sector pollutant estimates at the county level was employed. For these sectors, a national discharge estimate was produced by assuming that all plants employed "standard" production processes, generated similar amounts of waste per unit of output, and employed some "average" level of pollutant removal efficiency. The assignment of this national discharge estimate to individual counties was made in proportion to the county's employment in the industry. Estimates of employment by county were derived from the Public Use Sample of the 1970 Census (U.S. Department of Commerce, 1972).

Estimates of annual pollutant discharges from municipal sewage plants by county were derived from data on biochemical oxygen demand (BOD) and TSS discharges in a survey of 24,209 individual municipal sewage treatment plants (U.S. EPA, 1975).

The pollutants carried in nonpoint source discharges were also estimated for all nonpoint source sectors at the county level of detail. For certain sectors (for example, irrigation return flow, urban runoff, streambank erosion, and acid mine drainage) the analyses of other investigators were relied on, and their estimates of annual pollutant discharge accepted. For a second set of sectors (for example, ship ballast cleaning, accidental oil spills, and recreational boating) national estimates of discharges were prorated to counties using detailed information on the location and extent of each sector's activities. For a third set of sectors (for example, sediment from construction and mining), national estimates of discharge were prorated to counties according to the county's share of employment in these activities weighted by an estimate of runoff per acre.

A set of county-by-county estimates of sediment loss and sediment-related pollutant discharge from nonirrigated cropland, woodland, pasture land, and rangeland were derived from estimates of gross soil erosion made by the Department of Agriculture (for example, see USDA, 1973). The estimates were made by applying the Universal Soil Loss Equation to about 200,000 field sample points as part of the 1977 National Resource Inventory. Sediment delivery ratios and factors relating attached pollutants to sediment were then estimated for 156 geographically homogeneous regions. The ratios and factors were assumed to be constant for all counties within each region.

Translating Discharges into Measures of Ambient Water Quality

The county-by-county discharge estimates are linked to a detailed water quality network model. This network model consists of 304 rivers, 175 lakes and reservoirs and 37 bays. A total of 1,051 nodal points have been established along the rivers included in the model. Each county in the

United States is assigned to at least one node. (In certain cases more than one node was assigned if it was clear that the county's drainage went into more than one water body in the network.) The nodal points and water bodies are linked together in the network model in a logical downstream fashion.

The network model contains estimates for each river nodal point for the following hydrodynamic parameters:

Streamflow

Altitude

Width

Cross section area

Velocity

Surface water temperature

These estimates are drawn from published and unpublished data maintained by the U.S. Geological Survey.

Using the above data, pollution transport functions convert injections of pollutants at the nodal points to concentrations of pollutants at and between the nodes. Initial injections of pollutants at a node are mixed with quantities of pollutants arriving from upstream nodes; the mixture is diluted by the flow at the node; and the mixture is transported to the next node while certain constituents are removed on the way by natural processes. Dissolved oxygen (DO) concentrations are estimated using Streeter-Phelps relationships. A detailed discussion of the parameters and equations may be found in Gianessi, Peskin, and Young (1981). It should be noted that the model assumes that point source pollutants are inserted at specific nodal points. However, agricultural nonpoint pollutants are assumed to be evenly inserted along the reaches between the nodal points.

While the model in its current version endogenously calculates the dissolved oxygen content of each lake and river, some other water quality measures are fixed exogenously. The most important of these, from the perspective of reproduction and survival of the fish populations, are suspended solids, temperature, and pH.

Total suspended sediment loads (TSS) were not calculated from inventory discharges but were estimated for each node from a national map prepared by the U.S. Geological Survey in 1962 (Rainwater, 1962), showing average TSS concentrations in streams and rivers by hydrologic basin. The finest delineation of concentration groupings used in the map includes all water bodies with a TSS concentration less than 280 mg/liter (milligrams per liter or, equivalently, parts per million, ppm) in a single category. A search was made of stream monitoring data for 1972 for those nodal points that fall into basins in this TSS category (U.S. Geological Survey, 1972, 1973). When daily TSS measurements were available for these nodal points, and when the measurements indicated that a nodal point had measured TSS concentrations on average greater than 100 mg/liter, that point was not assigned to the category of nodal points with TSS concentrations less than 100 mg/liter. All other nodal points located in basins with a normal TSS concentration of less than 280 mg/liter (the USGS delineation) were assumed to fall into the less than 100 mg/liter category. All of the model's lakes and reservoirs were assumed to have average TSS concentrations of less than 100 mg/liter.

A map showing cold-water streams in the United States was prepared by Funk (1970). All of the model's nodal points (rivers and lakes) that fall within these cold-water stream areas were assumed to have a summer surface temperature of less than 18°C. In order to identify those nodal points whose surface water temperatures are as high as 32°C, a search was made of USGS ambient monitoring data (USGS, 1972, 1973; Briggs and Ficke, 1977). The maximum observed water temperature value was coded from the ambient data. As a result, there are nodal points whose high temperature is above 32°C. If there are no published temperature estimates for a nodal point, its summer temperature is assumed to be below 32°C.

A search was also undertaken of published ambient water quality data (U.S. Geological Survey, 1972, 1973; Briggs and Ficke, 1977) to obtain the ranges of pH measurements for nodal points. In this manner, some points were discovered to have pH measurements of less than 5.0. No points were discovered with observed pH measurements greater than 10.0. Special consideration was given to acid streams in Appalachia. All nodal points in river systems identified by the Appalachian Regional Commission (1969) as either "intermittently" or "continuously" affected by acid mine drainage were assigned a pH value of 4.0. All points with no observed pH measurements were assigned a neutral pH value of 7.0.

Translating Ambient Water Quality into Dominant Fish Populations (Fishery Types)

Prediction of values of classical water quality parameters still leaves us one large step from the desired goal of a water quality-related measure connected directly to recreational fishing. To make that step, the fishery biology and management literature was surveyed to produce a set of rules that appeared to capture whatever consensus existed on the water quality conditions, in terms of DO, pH, TSS, and (water) temperature necessary for the survival and reproduction of fish populations of various types (Nielsen, 1980).

A further assumption was also necessary, however, and it can be stated as follows:

If a water body can support a fish population dominated either by fish type j or type i, the state fishery manager will make sure that it is supporting whichever type is higher on a scale of preference broadly accepted by both fishermen and fishery management professionals. Our version of this scale is shown in table 2-1. It is based on research into angler preferences reported in Green and Wadsworth (1967) for Indiana, Brown (1973) for New York state, Volk and Montgomery (1973) for South Dakota, and Phillips and Ferguson (1977) for Wyoming; and on estimates of the values to anglers of fishing days of various types found in Charbonneau and Hay (1978). The major shortcoming of this particular version of an angler preference ordering may be in the northern bias of the studies just cited and the consequent handling of catfish, which are apparently highly prized as recreational fishing quarry in the southeastern United States, but which are included here in the lowest-valued group.[5]

The rules for determining the type of fishing supportable are summarized as logical if-then statements in table 2-2. From the logical statements we see that if dissolved oxygen is very low or temperature very high, the water will be unsuitable for recreational fishing. At higher dissolved oxygen levels, the fishery type depends on both temperature and suspended solids loadings. But because temperature is usually a function of natural factors, and suspended solids loadings are exogenously specified in our water quality model and do not change in response to FWPCA(A)/CWA implementation, policy-related water quality improvement, as reflected in changing species class, is a function of changing dissolved oxygen concentrations.[6] The improvement chains are shown in table 2-3.

TESTS OF DATA AND SYNTHETIC MODELING

The water quality network model is an ambitious effort which, to our knowledge, is the only model of its kind with such extensive coverage of U.S. lakes and waterways. However, this coverage is by no means complete,

Table 2-1. Preference Classes For Fish Species Sought By Freshwater Anglers

Broad species group	Assigned preference	Principal species	Secondary species
I. Rough fish	Low	carp freshwater drum buffalo bullhead catfish	paddlefish, smelt, fallfish, chub, goldfish, American eel, goldeye, mooneye, white sucker
II. Warm-water game fish-panfish	Medium	yellow perch white bass bluegill crappie rock bass pumpkinseed lake herring northern pike	white perch, warmouth, orangespotted sunfish, green sunfish, longear sunfish, chain pickerel, grass pickerel
	High (warm water/ cool water)	largemouth bass smallmouth bass walleye striped bass	spotted bass, sauger, muskellunge
III. Cold-water game fish	High (cold water)	rainbow trout brook trout brown trout lake trout	cutthroat trout, Atlantic salmon, steelhead, Pacific salmon(s)

Source: Larry Nielsen, 1980. "Water Criteria and Angler Preference for Important Recreational Fishes." EPA Benefits Project Recreation Working Paper #3. Unpublished. (Washington, D.C., Resources for the Future).

as shown in table 2.4. Although the model covers about 40 percent of the total national acreage of fishable water, coverage in some eastern states (Maine, Maryland, Massachusetts, Delaware, New Hampshire) is very spotty and coverage for one of the most populous states, New York, is only 20 percent. For several of the eastern states, lakes and reservoirs are not represented in the model.

Furthermore, the results of the model have not been systematically verified, so at present users cannot be sure of the accuracy of its dissolved oxygen predictions. The model is fairly crude in comparison with more detailed models built for single river basins. It contains very long

Table 2-2. Logical Statements Translating Ambient Water Quality Measures into Three Fishery Groups According to Water Quality

Rule No.			Water Class
1.	If pH is above 10.0 then	:	No fishery supportable
2.	If pH is below 5.0 then	:	No fishery supportable
3.	If D.O. is below 2.0 mg/liter then	:	No fishery supportable
4.	If temp. is above 34°C then	:	No fishery supportable
5.	If summer streamflow is zero then	:	No fishery supportable
6.	If D.O. is between 2.0 mg/liter and 3.0 mg/liter then	:	Rough fish
7.	If D.O. is above 3.0 mg/liter and temp. is above 32°C then	:	Rough fish
8.	If D.O. is above 3.0 mg/liter and TSS is greater than 100 mg/liter then	:	Rough fish
9.	If D.O. is greater than 3.0 mg/liter and temp. is between 18°C and 32°C and TSS is below 100 mg/liter then	:	Warm-water game fish/ panfish
10.	If D.O. is greater than 5.0 mg/liter, temp. is less than 18°C and TSS is less than 100 mg/liter then then	:	Cold-water game fish

Key: D.O.: 90-day average summer dissolved oxygen concentration
Temp: 90-day average summer temperature
TSS: Annual average total suspended solids

Table 2-3. Fishery Types as Determined by Water Quality Using Nielsen's Rules

Water quality in terms of dissolved oxygen	Exogenously determined water quality conditions: Cold water temperature ≤ 18°C — Low solids water TSS ≤ 100 mg/liter	Cold water temperature ≤ 18°C — High solids water TSS > 100 mg/liter	Warm water temperature between 18°C and 32°C — Low solids water TSS ≤ 100 mg/liter	Warm water temperature between 18°C and 32°C — High solids water TSS > 100 mg/liter
Dissolved oxygen				
>5.0 mg/liter	Cold-water game fish			
≥3.0 mg/liter			Warm-water game fish/ panfish	
>2.0 mg/liter	Rough fish	Rough fish	Rough fish	Rough fish
<2.0 mg/liter	Unfishable[a]	Unfishable[a]	Unfishable[a]	Unfishable[a]

[a]Other unfishable conditions include pH greater than 10 or less than 5; temperature greater than 34°C; summer flow zero. Between 32° and 34°C the water is assigned to rough fish.

Table 2-4. Approximate Percent of Fishable Water Covered in the RFF Water Quality Network Model

State	U.S. Department of Interior actual (10^3 acres)	RFF model (10^3 acres)	% Coverage by model
ALABAMA	574	350	61
ARIZONA	200	127	64
ARKANSAS	811	354	44
CALIFORNIA	863	394	46
COLORADO	176	30	17
CONNECTICUT	48	10	21
DELAWARE	5	0.2	4
DISTRICT OF COLUMBIA	5	0.6	12
FLORIDA	1968	586	30
GEORGIA	625	288	46
IDAHO	570	261	46
ILLINOIS	371	154	48
INDIANA	129	91	71
IOWA	335	70	21
KANSAS	301	128	43
KENTUCKY	526	342	65
LOUISIANA	2215	691	31
MAINE	1638	28	2
MARYLAND	151	6	4
MASSACHUSETTS	133	7	5
MICHIGAN	940	326	35
MINNESOTA	2594	1517	58
MISSISSIPPI	551	224	41
MISSOURI	697	283	41
MONTANA	1015	573	56
NEBRASKA	176	106	60
NEVADA	351	286	81
NEW HAMPSHIRE	197	8	4
NEW JERSEY	119	16	13
NEW MEXICO	135	84	62
NEW YORK	1735	353	20
NORTH CAROLINA	822	99	12
NORTH DAKOTA	541	505	93
OHIO	244	54	22
OKLAHOMA	904	374	41
OREGON	592	178	30
PENNSYLVANIA	122	94	77
RHODE ISLAND	n.a.	n.a.	n.a.[a]
SOUTH CAROLINA	689	221	32
SOUTH DAKOTA	723	407	56
TENNESSEE	663	494	75
TEXAS	1469	908	62
UTAH	372	278	75
VERMONT	139	161	116[b]
VIRGINIA	452	106	23
WASHINGTON	1158	248	21
WEST VIRGINIA	141	37	26
WISCONSIN	1138	254	22
WYOMING	414	118	29
TOTAL	30615	12127	40[c]

Source: For column 1.

Actual: U.S. Department of the Interior, Fish and Wildlife Service. 1968. National Survey of Needs for Hatchery Fish, Bureau of Sport Fisheries and Wildlife Publication No. 63, Washington, D.C.

[a]Not covered in RFF WQNM.

[b]Exceeds 100 due to allocation of Lake Champlain

[c]Weighted average.

river segments (66 miles on average); combines the discharges of all point sources in a county and assumes they enter the water network at only one point per county; assumes that land runoff is uniformly distributed along major waterways when in fact much runoff enters major waterways through tributaries at point locations; and neglects algal production and decay, a major source of BOD. Thus, on a priori grounds we should be cautious in using the model's predictions as the basis for distributing fishable water across species classes.[7]

Because we gathered survey data on the pre-policy (1972) proportions of fishable water and the allocation of that water among fishery types, we can check the accuracy of our modeling route to the same information. That is, we can see how well our data on discharges before FWPCA(A)/CWA (and on nonmodeled ambient conditions), in combination with our model of water quality and fishery type, reproduce the observations of our expert respondents in the states. Comparisons of this sort are summarized in tables 2-5 through 2-8, where we show data for the pre-policy base period.

Table 2-5 - the percentage of fishable water by state from our survey and from application of the water quality network model.

Table 2-6 - comparisons of the means and correlation coefficients for the data in table 2-5.

Table 2-7 - the breakdown of fishable water by state into the three major fishery types from our survey, from the data plus model, and from the data plus model adjusted using the <u>Survey of Hatchery Fish</u>.[8]

Table 2-8 - comparisons of the means and correlation coefficients for the data in table 2-7.

The message of these comparisons is that the modeling results are close to those from the survey--perhaps remarkably close when we consider the many possible sources of difference. These include:

errors in the discharge inventory

the highly aggregated and approximate character of the water quality model

the rough nature of the information on TSS and summer water temperatures

the rules of thumb used to translate water quality measures into fishery types (and to apportion water between fishable and nonfishable categories)

Table 2-5 Percentage of Fishable Water: Two Sources

State	RFF Survey	RFF WQNM
ALABAMA	99	96
ARIZONA	97	96.7
ARKANSAS	85.5[a]	99.9
CALIFORNIA	85	77.7
COLORADO	95	99.9
CONNECTICUT	85	92.1
DELAWARE	99	99
DISTRICT OF COLUMBIA	n.a.	100
FLORIDA	99	99.2
GEORGIA	97	97.6
IDAHO	95	97.5
ILLINOIS	98	91.3
INDIANA	99	98.3
IOWA	99	66.2
KANSAS	n.a.	99.7
KENTUCKY	79	98.0
LOUISIANA	92.5[a]	100.0
MAINE	90	100.0
MARYLAND	n.a.	100.0
MASSACHUSETTS	n.a.	100.0
MICHIGAN	99	98.0
MINNESOTA	82.4	99.5
MISSISSIPPI	100	100
MISSOURI	98	99.6
MONTANA	99.6	98.6
NEBRASKA	95	100
NEVADA	96	100
NEW HAMPSHIRE	98	100
NEW JERSEY	97	96.9
NEW MEXICO	100	97.8
NEW YORK	97.5[a]	95.1
NORTH CAROLINA	n.a.	98.4
NORTH DAKOTA	90	100
OHIO	95	91.9
OKLAHOMA	100	98.3
OREGON	93.5[a]	100
PENNSYLVANIA	79[a]	64.7
RHODE ISLAND	95	n.a.
SOUTH CAROLINA	87.5[a]	100
SOUTH DAKOTA	90[a]	94.5
TENNESSEE	100	100
TEXAS	97.8[a]	97.0
UTAH	96.8	100
VERMONT	100	100
VIRGINIA	93.5[a]	100
WASHINGTON	95	100
WEST VIRGINIA	20	67.1
WISCONSIN	n.a.	93.4
WYOMING	95	100
UNWEIGHTED AVERAGE	92.6	95.8

[a] The figures reported are the average of two or more state reports:

	Report 1	Report 2	Report 3
Arkansas	74	97	...
Louisiana	90	95	...
New York	95	100	...
Oregon	87	100	...
Pennsylvania	75	77	85
South Carolina	80	95	...
South Dakota	80	99.9	...
Texas	96.6	99	...
Virginia	92	95	...

Table 2-6. Percent Fishable Water: Comparison of Means and Correlations

	RFF Survey		RFF WQNM
UNWEIGHTED			
Mean	92.6 (N = 43)		95.8 (N = 48)
Mean difference (Survey - RFF WQNM)		-2.9 (N = 42)	
t, Mean difference		1.73	
correlation coefficient		0.56*	
ACREAGE WEIGHTED			
Mean	93.92 (N = 43)		97.18 (N = 48)
Mean Difference (Survey - RFF WQNM)		-4.03 (N = 42)	
t, Mean difference		3.09*	
correlation coefficient		0.29	

*Significant at 5 percent level.

Table 2-7. Alternative Allocations of Precontrol Fishable Water Acreage by Species Group

State	Fractions of state fishable water by class of fishing								
	Cold-water game fish			Warm-water game fish/panfish			Warm-water rough fish		
	RFF Survey[a]	RFF WQNM SHF[b]	RFF WQNM[c]	RFF Survey	RFF WQNM SHF	RFF WQNM	RFF Survey	RFF WQNM SHF	RFF WQNM
ALABAMA	.01	0	0	.95	1.00	1.00	.04	0	0
ARIZONA	.35	.24	.22	.55	.74	.76	.10	.02	.02
ARKANSAS	.02	.04	.06	.74	.78	.76	.24	.18	.18
CALIFORNIA	.25	.44	.87	.75	.34	.08	0	.22	.05
COLORADO	.75	.64	.40	.20	.31	.51	0	.05	.09
CONNECTICUT	.40	.31	.46	.60	.69	.54	0	0	0
DELAWARE	0	0	0	.96	.96	.96	.04	.04	.04
DISTRICT OF COLUMBIA	n.a.	0	0	n.a.	0	0	n.a	1.00	1.00
FLORIDA	0	0	0	.90	1.00	1.00	.10	0	0
GEORGIA	.05	.12	0	.92	.84	.96	.03	.04	.04
IDAHO	.75	.89	.63	.25	.10	.33	0	.01	.04
ILLINOIS	.01	0	0	.84	.98	.98	.15	.02	.02
INDIANA	.02	.03	.01	.94	.31	.32	.04	.66	.67
IOWA	.02	0	0	.93	.92	.92	.05	.08	.08
KANSAS	n.a.	0	0	n.a.	.96	.96	n.a.	.04	.04
KENTUCKY	.02	.07	.02	.77	.72	.76	.21	.21	.22
LOUISIANA	0	0	0	.80	.95	.95	.20	.05	.05
MAINE	.80	.57	.72	.20	.43	.28	0	0	0
MARYLAND	n.a.	0	0	n.a.	.77	.77	n.a.	.23	.23
MASSACHUSETTS	n.a.	.53	.10	n.a.	0	0	n.a.	.47	.90
MICHIGAN	.56	.17	.97	.43	.28	.01	.01	.55	.02
MINNESOTA	.04	.06	.01	.79	.94	.99	.17	0	0
MISSISSIPPI	0	0	0	.90	.82	.82	.10	.18	.18
MISSOURI	0	.07	.01	.71	.91	.97	.29	.02	.02
MONTANA	.54	.71	.43	.13	.28	.55	.33	.01	.02
NEBRASKA	.05	.23	0	.35	.66	.86	.60	.11	.14
NEVADA	.59	.60	.74	.40	.40	.26	.01	0	0
NEW HAMPSHIRE	.20	.58	80	.80	0	0	0	.42	.20
NEW JERSEY	.16	.26	.16	.76	.41	.47	.08	.33	.37
NEW MEXICO	.40	.28	.24	.60	.68	.72	0	.04	.04
NEW YORK	.20	.21	.44	.62	.58	.41	.18	.21	.15
NORTH CAROLINA	n.a.	.03	.06	n.a.	.96	.93	n.a.	.01	.01
NORTH DAKOTA	.30	.08	0	.43	.92	1.00	.27	0	0
OHIO	.01	0	0	.85	.28	.28	.14	.72	.72
OKLAHOMA	0	0	0	.90	1.00	1.00	.10	0	0
OREGON	.90	.97	.94	.10	.02	.04	0	.01	.02
PENNSYLVANIA	.18	.28	.15	.78	.66	.78	.04	.06	.07
RHODE ISLAND	.05	n.a.	n.a.	.90	n.a.	n.a.	.05	n.a.	n.a.
SOUTH CAROLINA	.05	.01	.01	.90	.97	.97	.05	.02	.02
SOUTH DAKOTA	.08	0	0	.75	.96	.96	.17	.04	.04
TENNESSEE	n.a.	.17	.01	n.a.	.79	.94	n.a.	.04	.05
TEXAS	0	0	0	.80	.98	.98	.20	.02	.02
UTAH	.58	.27	.04	.36	.71	.93	.06	.02	.03
VERMONT	.33	.29	.32	.65	.71	.68	.02	0	0
VIRGINIA	.07	.07	.07	.85	.90	.90	.08	.03	.03
WASHINGTON	.50	.82	1.00	.25	.09	0	.25	.09	0
WEST VIRGINIA	.05	.04	.23	.30	.39	.31	.65	.57	.46
WISCONSIN	.02	.03	.02	.84	.93	.94	.14	.04	.04
WYOMING	n.a.	.96	.82	n.a.	.04	.17	n.a.	0	.01

[a] RFF Survey is the RFF Survey of Fishable water conducted by mail and directed at State fish and game agency officials.

[b] RFF Water Quality Network Model as modified by data from U.S. Department of Interior, Fish & Wildlife Service, 1968 National Survey of Needs for Hatchery Fish. Bureau of Sport Fisheries and Wildlife, Pub. No. 63 (Washington, D.C.).

[c] RFF Water Quality Network Model.

Table 2-8. Comparison of Allocations of Fishable Water by Species Group: Means and Correlation Coefficients

A. Mean share of total fishable water by class[a]	Alternative Sources		
	RFF Survey	RFF WQNM SHF	RFF WQNM
Unweighted: all available states			
Cold-water game fish	0.24	0.23	0.23
Warm-water game fish/panfish	0.63	0.63	0.64
Warm-water rough fish	0.12	0.14	0.13
Number of states	42	48	48
Weighted by state freshwater: 41 comparable states			
Cold-water game fish	0.27	0.21	0.25
Warm-water game fish/panfish	0.60	0.71	0.70
Warm-water rough fish	0.13	0.08	0.05
Weighted by state fisherman population: 41 comparable states			
Cold-water game fish	0.21	0.18	0.26
Warm-water game fish/panfish	0.68	0.65	0.62
Warm-water rough fish	0.11	0.17	0.12

B. Simple correlation coefficients: Survey vs. two versions of model[b]

	RFF WQNM SHF	RFF WQNM
Unweighted: All available states		
Cold-water game fish	0.75*	0.65*
Warm-water game fish/pan fish	0.52*	0.42*
Warm-water rough fish	0.17	0.23
Weighted by state freshwater: 41 comparable states		
Cold-water game fish	0.61*	0.57*
Warm-water game fish/panfish	0.55*	0.47*
Warm-water rough fish	0.00	0.19
Weighted by state fisherman population: 41 comparable states		
Cold-water game fish	0.55*	0.53*
Warm-water game fish/panfish	0.33*	0.30*
Warm-water rough fish	-0.07	0.09

[a]Mean shares across classes from any particular source may not add to 1.00 due to rounding.

[b]An asterisk (*) indicates significance at the 5 percent level.

the substantially less than complete coverage in the water quality network model of the fishable water in the United States

A more careful assessment of the differences between model and survey results would have to conclude that the WQNM is considerably better at giving national shares of fishable water by fishery type than at producing state-by-state estimates. This is illustrated by the contrast between means and correlations in table 2-8. In part A of the table a comparison of unweighted means and a comparison of means weighted by fisherman population both indicate a fairly close correspondence across the sources in fishable water allocations for the average state, while the comparisons weighted by water acreage suggest some discrepancies between the <u>RFF Survey of Fishable Water</u> and the water quality network model in the two warm-water classes.

Though there is reasonable agreement on average across the states, there are considerable differences for particular states, as is revealed by part B of table 2-8. There, we report the unweighted and weighted simple Pearsonian bivariate correlation coefficients between the RFF survey allocations by species group and the allocations obtained from two versions of the water quality model (raw and adjusted by <u>Hatchery Fish</u> data). These correlations show that there is a strong positive linear relationship between the RFF survey's cold-water allocation and each of the other sources' cold-water allocations. The relationship is weaker for the warm-water game fish-panfish allocations and, unfortunately, very weak (indeed statistically insignificant) for the rough fish allocations.

A similar but probably less severe problem appears when the model is compared with the state survey reports on the proportion of <u>total available</u> water which is classified as fishable. Table 2-5 displays the data and table 2-6 shows that there is a reasonable correspondence between the two sources as measured by their unweighted means and correlation coefficient. However, on an acreage-weighted basis, the two sources are significantly different on average and observation-by-observation.

This overall result is not unexpected because our survey measures of the percentage of total water not fishable are more all-inclusive than the model's. Some water is not fishable because it is naturally too shallow, intermittent or brackish; is subject to winter freezeout; or is closed due to chemical contamination (PCB's, DDT, mercury, Kepone). All of these factors are reflected in the state reports, but not in the model. This explains the direction of the difference in the mean percentage of fishable water between the network model and the state reports.

PROJECTING THE EFFECTS OF IMPLEMENTATION

Implementation of existing national water quality legislation will affect pollution discharges, which in turn will affect measures of water quality and hence the availability of fishable water in total and of higher quality fisheries in particular. Once the discharges have been changed, the water quality predictions and fishery-type allocations are straightforward applications of the methods already discussed. But our consideration of the difference between model and reality suggests that at least some allowance should be made for coverage imperfections. Both the alteration of discharges to reflect implementation of the FWPCA(A)/CWA and the process of adjusting our projections will be described in this section.

How Discharges Are Changed to Reflect Implementation

The industrial point source categories in the model correspond to the subcategories established by EPA for development of effluent regulations. By substituting the BOD discharge limits required by the Best Practicable Treatment (BPT) regulations in each subcategory for the 1972 discharge coefficients (raw loads net of in-place removal capability) the post-BPT BOD discharges from industrial point sources may be calculated straightforwardly, assuming perfect industry compliance with the regulations. A similar approach was undertaken for municipal sewage plants, where all plants are assumed to be upgraded to secondary treatment (ST).

A second policy simulation combines the BPT/ST policy discharge estimate with an assumed nonpoint-source control policy--one that reduces sediment discharges from nonirrigated cropland by 100 percent. This is translated into BOD discharge reduction using the rule of thumb that each pound of organic carbon in nonpoint source sediment is equivalent to 0.1 pound of biochemical oxygen demand.[9] No change in ambient levels of TSS is assumed after this discharge reduction.[10]

Other changes in water quality attributable to implementation of FWPCA(A)/CWA must, at this stage of the model's development, be simulated by operating directly on the ambient quality measures. Thus, for acid mine drainage control it is assumed that in all streams and lakes affected by such drainage the pH level is raised to at least 5.0 everywhere.[11] To mimic the operation of the highest stage of point source discharge control, Best Available Treatment (BAT), we simply made use of the link between BAT and stream quality implied by the FWPCA(A)/CWA language and assumed that under BAT no point in the water network could have a dissolved oxygen level lower than 5.0 ppm.[12]

Projecting Fishable Water in Total and by Type of Dominant Population

Having predicted the effect of policy implementation on discharges, we had next to project the resulting ambient water quality and the consequent shifts in quantity and distribution of fishable water. The basis for these projections was the water quality network model, and the simplest approach would have been to use the model's projections of fishable acres by dominant fish species class to obtain ratios of projected to baseline acres by class. The total projected fishable water acreage would then be found as the sum over the fishery classes. We felt, however, that because one of our major interests was the change in fishable water in total, it would be better not to leave that projection as an uncontrolled result of three separate calculations because the model did not seem to perform very well for the rough fish proportion of fishable water. Rather, we began by projecting the change in total fishable acres, using the baseline data from *Hatchery Fish* and the model's results for post-policy conditions. We then projected the distribution of this acreage over fishery classes using the model's results suitably adjusted to maintain the desired total. This adjusted projection technique can be seen most efficiently with the aid of some simple algebra. For this purpose, we shall define for a particular state:

Subscript t = 0 for precontrol conditions
= 1, 2, 3 ... postcontrol conditions*

(*There are several different postcontrol conditions examined, as we have just described, but as the forecasting technique does not differ among them it is not necessary here to distinguish them.)

F_0 = Total fishable water acreage in the state in precontrol conditions, based on *Hatchery Fish*

F_t = Forecast of total acres of fishable water, postcontrol

G_t = Total fishable water acreage in the state postcontrol, based on RFF WQNM and the rules in table 2-2

α_{j0} = Proportion of fishable water in state allocated to fishery type j in period 0 on basis of RFF Survey

$$j \begin{cases} = T \text{ for cold-water game} \\ = B \text{ for warm-water game} \\ = R \text{ for rough} \end{cases}$$

m_{jt} = Proportion of fishable water in state allocated to fishery type j in period t on basis of RFF WQNM and Nielsen's rules

Using this notation we can describe our forecasting methods as follows:

I. Forecast of the post-policy acreage of fishable water:

$$F_t = F_0 \cdot \frac{G_t}{G_0}$$

II. Forecast of the post-policy acreage of fishable water available for fishery type j:

$$\text{Define } q_{jt} \equiv \frac{m_{jt}\ G_t}{m_{j0}\ G_0}$$

or the ratio of fishery areas in pre- and post-control conditions, according to RFF WQNM. Then forecast the area of fishery type j in period t as:

$$F_{jt} \equiv q_{jt} \left(\alpha_{j0}F_0\right) \left[\frac{F_t}{\sum_j \left(q_{jt}\alpha_{j0}F_0\right)}\right]$$

where $\alpha_{j0}F_0$ is the baseline acreage for this fishery type, and the last term on the right is an adjustment factor guaranteeing that the sum of the forecast F_{jt} will be equal to F_t (i.e., $F_t = \sum_j F_{jt}$).

III. Then the forecast fishery <u>share</u> of fishable water of type j is given by:

$$X_{jt} \equiv \frac{F_{jt}}{F_o \frac{G_t}{G_0}} = \frac{F_{jt}}{F_t} \text{ and } \sum_j X_{jt} = 1$$

PROJECTIONS OF CHANGES IN TOTAL FISHABLE ACREAGE AND CHANGES IN SHARES OF TOTAL FISHABLE ACREAGE BY FISHERY TYPE WITH IMPLEMENTATION OF FWPCA(A)/CWA

Our projection of the change in the quantity of fishable water (by fishery type) in reaction to implementation of FWPCA(A)/CWA is the driving force behind our estimate of dollar benefits since this change is translated into a change in the amount of participation and its allocation across fishery classes via the equations estimated in chapters 3 and 4.

These equations can either be evaluated at the national level by using the aggregate national changes in fishable water and fishable water allocations across fishery type predicted by the network model; or at the disaggregated state level by using the model's state-level predictions, whereupon national changes would be the sum of changes across states. The nonlinear nature of the logit equations estimated in chapters 3 and 4 argues for evaluating the effects separately for each state (that is, 49 separate evaluations of the estimated equations at the state means for all explanatory variables except fishable water or its allocation across fisheries) and then summing. However, as just discussed, it seems apparent that the water quality network model, at least in our application, gives a much more reliable picture on average than it does for any particular state.

It is therefore our opinion that projections of changes in total fishable water and its allocation across fisheries groups are best done using the model's aggregate level, and not built up state by state.
The specific steps in our procedure are to:

> Calculate the percentage increase from the baseline in total fishable water represented in the water quality network model under the following "scenarios":
>
> 1. Best Practicable Control Technology Currently Achievable (BPT)
> 2. BPT with Cropland Sediment Control
> 3. BPT and Cropland Sediment Control and pH > 5.0 (from mine drainage control)
> 4. All watercourses to have at least 5.0 ppm dissolved oxygen content (a proxy for Best Available Technology Economically Achievable)

> Use the water quality network model national allocations of fishable water by species class, pre- and post-policy, to project the shares of the total fishable water accounted for by game fisheries (cold and warm water) and rough fisheries (usually warm but occasionally cold water). The algebra for this step is as described above, substituting "nation" for "state."

The results for the nation as a whole of projecting the effects of the four scenarios using the water quality network model and the adjustments for its less than complete coverage are summarized in table 2-9. For completeness, we show both the numbers coming directly out of the model runs (the first four columns) and the projections after correction for the model's incomplete coverage (columns 5 through 8). These last four columns contain the information used in the calculation of increased participation, and hence benefits. With perhaps one exception these results conform to expectations: as pollution control proceeds to higher levels total fishable acreage increases and so do the shares of game fisheries relative to rough fisheries.[13]

On the other hand, the reader may be struck by the small size of these changes. For example, the total change in fishable acreage from pre-FPWCA(A)/CWA conditions to scenario 4 (roughly BAT) is only 2.06 percent. The share of trout water in the total increases by only 15 percent and that of bass water by 12 percent, while the share of rough fish water falls by 87 percent. (All these percentages reflect the adjustment, described above, that is required to maintain total acreage.) The percentage changes taken directly from the model are somewhat different, with trout water increasing by 2 percent, bass by 5 percent, and rough declining by 88 percent. The changes seem small because in the base period fishable water, and game fishing specifically, were extensively available. We were not starting from a situation of severe lack of fishable water. The incremental improvements represented by the scenarios do, however, show up as very large declines in the nationwide quantity of <u>un</u>fishable water. Between the base case and scenario 4, unfishable acreage declines from 2.76 percent of the total of all water, to 0.76 percent, or by over 70 percent. All the relative changes are summarized in table 2-10.

Table 2-9. Summary of Results of Projections Using Water Quality Network Model for Pollution Control Scenarios

	WATER QUALITY NETWORK MODEL				BASELINE AND PROJECTED			
	Baseline and projected areas	Baseline and projected fishery shares			Fishable water total	Fishery shares[a]		
	G_0, G_t (1,000 acres)	m_{T0}, m_{Tt} cold water game	m_{B0}, m_{Bt} warm water game	m_{R0}, m_{Rt} rough	F_0, F_t (1,000 acres)	α_{T0}, X_{Tt} cold water game	α_{B0}, X_{Bt} warm water game	α_{R0}, X_{Rt} rough
Pre FWPCA(A)/CWA conditions	12,015	0.167	0.784	0.049	30,615	.204	.684	.112
Pollution control scenarios								
1. BPT	12,057	0.168	0.817	0.015	30,721	.227	.736	.037
2. BPT + cropland sediment control	12,140	0.167	0.822	0.011	30,934	.231	.742	.027
3. BPT and sediment control + pH control	12,219	0.169	0.820	0.011	31,134	.229	.743	.028
4. All water courses at $\geq$ 5.0 mg/liter O_2 content	12,263	0.170	0.824	0.006	31,247	.232	.755	.013

[a]Baseline shares of fishable water by fishery type from table 2-8 (RFF Survey, weighted averages, weighted by state fishermen).

Table 2-10. Percentage Changes from Baseline Situation: Unfishable and Fishable Water, and Fishery Shares

	Percentage changes from base in				
Scenario	Quantity of unfishable water	Percentage fishable water	Shares of fisheries[a] Trout	Bass	Rough
1. BPT	-12.2	0.3	0.6	4.2	-69.4
2. BAT + cropland sediment control	-36.7	1.0	0.0	4.8	-77.6
3. BPT, sediment control and pH control	-59.6	1.7	1.2	4.6	-77.6
4. Dissolved oxygen concentrations everywhere at least 5 ppm	-72.6	2.1	1.8	5.1	-87.8

[a]Comparison of unadjusted model results.

CONCLUSIONS

We have devoted a good deal of space to modeling the ambient effects of environmental regulations because it is apparently not widely understood, at least among economists, lawmakers, and administrators, that the difficulty in measuring the benefits of an environmental program is not to be found entirely in valuation questions. The logically prior matter of estimating the impact the law will have on those features of the ambient environment valued by people is itself very difficult. Sometimes even the baseline data from which to measure the effect to be ascribed to the law or regulation are lacking. Beyond that, often there are no well-developed models of the links between law and ambient effect; especially of the effect of the law on discharges (or analogous human actions having an impact on the environment), and of the effect of the "discharges" on meaningful measures of ambient quality. Finally, even if we have all the pieces, it is a far from trivial matter to put them together in a suitably comprehensive framework for the assessment of laws or regulations with national scope.

The attempt here to show how the implementation of FWPCA(A)/CWA and certain related policies will change the amount and species-class mix of fishable water in the United States suffers to some extent from each of these infirmities: imperfect baseline data, modeling requirements pushing to and beyond the current limits of our ability, and imperfect coverage of total U.S. watercourses. Nonetheless, we believe this effort demonstrates that some progress is possible.

NOTES

1. Because of the extreme differences between the lower forty-eight states and Alaska and Hawaii, and because of the dominance of the former in population terms, we decided to concentrate on the contiguous forty-eight states. Hence, "United States" in this book means the forty-eight states of the mainland and the District of Columbia.

2. Our communications with state fish and game officials revealed that it was virtually impossible for many of the respondents to identify (or allocate total acreage to) water bodies which would support panfish but not warm-water game fish, since these species groups so frequently occur together in "ecologically balanced" water bodies. Similar difficulties were encountered with our distinction between nonsporting food fish and rough fish.

3. Even within a given state, there were often discrepancies in those instances where we received more than one return. In such cases, unweighted averages of the reported allocations were used. We emphasize that the returns of this survey do not reflect the official estimates of state fish and game commissions, but are merely unofficial "guesstimates."

4. The entire exercise holds the level of industrial output and the size of the total U.S. population constant at their base (pre-FWPCA(A)/CWA) levels.

5. Generalizations about preferences are subject to two caveats. The first is that preference sometimes is equated with use. Thus, species fished for are assumed to represent the desires of the public. But studies have indicated discrepancies between species preferred and those most frequently fished for (Brown, 1973; Volk and Montgomery, 1973) and people have been shown to express desires to fish for other species which have better fighting qualities, taste better, or are more fun to catch (Duttwiler, 1974). Second, as observed in the text, species preferences seem to vary across regions, perhaps because of regional availability and an associated learning-by-doing effect. This categorization does not do justice to the variety of fish species potentially available across the United States; nor does it capture within-category variation (for example, brown trout are more tolerant of adverse conditions than are brook or rainbow trout); but it does represent the finest breakdown for which other relevant data were available.

6. Other, similarly rough, rules for translating physical-chemical water quality parameters into measures of fishing quality may be found elsewhere. For example, U.S. EPA in "The Red Book" (1976) asserts that "a minimum concentration of dissolved oxygen to maintain good fish population is 5.0 mg/liter. The criterion fo. salmonid spawning beds is a minimum of 5.0 mg/liter in the interstitial water of the gravel" (p. 123). Alabaster (1969) summarized the then current state of knowledge about differential sensitivity of fish species to pollution and concluded:

> It seems that there is already sufficient information from laboratory and field observations to conclude that coarse (rough) fish are generally somewhat less sensitive to pollution than trout, though differences between species are not always the same for all poisons. (Authors' insertion).

More recently, the English Water Research Centre summarized and commented on the development of European water quality standards designed to protect fish life (with some species differentiation). (Water Research Centre, 1979.) This report emphasized that the continuing gaps in knowledge of fish response to pollution made guidelines more desirable than inflexible requirements.

7. Initial runs of the model were checked by the authors and our consultant, Professor Larry Nielsen, against published surveys of fish populations (O'Dell and coauthors, 1976; Ohio River Valley Sanitation Commission, 1978; U.S. EPA, 1980; Wisconsin Department of Natural Resources, 1978) and our own Survey of Fishable Water. Where major discrepancies between the model's predictions and observed reality were discovered, the model was corrected by substituting observed for predicted conditions.

8. Specifically, from the Survey of Hatchery Fish we can obtain the fraction of state fishable water which is suitable for cold-water game fish, C, and the fraction suitable for warm-water fish of all types, W (i.e., W = 1-C). From the WQNM we produce the fraction of all fishable water covered in the model, warm and cold, suitable for warm-water game fish-panfish, g, and the fraction suitable for warm-water food fish-rough fish, r. Then, our Hatchery Fish-based estimates of the fractions of cold-water game fish (c), warm-water game fish-panfish (g') and warm-water food fish-rough fish (r') are:

$$c = C$$

$$g' = W(g/g+r)$$

$$r' = W(r/g+r)$$

9. Control of soil erosion is not part of FWPCA(A)/CWA, but is a critical ingredient in any overall program with a "fishable" water goal. Fred W. Johnson, Water Resources Coordinator of the Pennsylvania Fish Commission, said in response to our Fishable Water Survey, "Soil erosion is viewed by this agency as the most widespread current form of pollution." The organic carbon content of soil runoff was estimated regionally from USDA data (e.g., USDA, 1973).

10. Reliable models of sediment transport and deposition are unavailable. It might be argued that if the modeled rivers were in sediment equilibrium prior to sediment control, the imposition of sediment control might lead to channel diversion and scouring, which could restore the former equilibrium with no change in sediment loads.

11. Again quoting Fred Johnson of the Pennsylvania Fish Commission:

The principal cause of unfishable water in Pennsylvania today is abandoned coal mines. It will probably be useful to know that in 1970, we estimated that 73% of the unfishable stream acreage in the state was the result of coal mining activity. Approximately 3,000 miles of major streams were affected by this single activity. In 1985 this most pervasive source is expected to account for 92% of streams remaining unfishable as a result of pollution caused by sources covered by the provision of PL 92-500 (FWPCA(A)). As can be clearly seen, other sources are expected to be virtually eliminated, but an estimated 2,300 miles of virtually dead water resulting from mine drainage and sedimentation will remain.

12. This link is implied by: (1) the original timing of the BAT requirement, to be achieved by July 1, 1983, or simultaneously with the requirement that the interim water quality goal ("provides for the protection and propagation of shellfish and wildlife and provides for recreation in and on the water": (Sec 101(a) (2)); and (2) the requirement that if BAT will not allow attainment of that quality goal, stricter discharge limits may be imposed by the administrator of EPA (Sec 302(a)). For a discussion of the water quality criteria which are expected to reach the goal, see EPA (1976), David, and coauthors, (1979), and Council on Wage and Price Stability (1979).

13. There is a small decline in the cold-water game fishery share between scenarios 2 and 3, while the overall warm plus cold-water game fishery share stays constant. This is simply a result of the character of the acreage added to fishable water by pH control. In the model's results, the decline occurs in the warm-water game fishing share.

REFERENCES

Alabaster, J.S. 1969. "The Comparative Sensitivity of Coarse Fish and Trout to Pollution," in Proceedings of the Fourth British Coarse Fish Conference (Liverpool University).

Appalachian Regional Commission. 1969. Acid Mine Drainage in Appalachia (Washington, D.C.)

Briggs, John C. and John F. Ficke. 1977. Quality of Rivers of the U.S., 1975 Water Year--Based on the National Stream Quality Accounting Network (NASQAN). USGS Open-File Report 78-200, May (Reston, Virginia, U.S. Department of the Interior, Geological Survey).

Brown, T. L. 1973. The 1973 New York Statewide Angler Study (Ithaca, N.Y., Cornell Univ., Dept. of Natural Resources) 117 pp.

Charbonneau, John and Michael J. Hay. 1978. "Determinants and Economic Values of Hunting and Fishing." A paper presented at the 43rd North American Wildlife and Natural Resources Conference, Phoenix, Arizona, March 18-22, 20 pp.

Council on Wage and Price Stability, Executive Office of the President. 1979. "Comments Before the Environmental Protection Agency on Water Quality Standards, Surface Waters of the State of Ohio." (Washington, D.C., October 19) 21 pp.

David, J. C., G. I. Bresnick, P. Doudoroff, T. R. Doyle, A. J. Mearns, J. B. Pearce, J. T. Peterka, J. G. Robinson and D. L. Swanson. 1979. "Dissolved Oxygen," in A Review of the EPA Red Book: Quality Criteria for Water (Bethesda, Md., American Fisheries Society) pp. 169-174.

Duttwiler, M. W. 1974. "Use of Questionnaire Surveys in Forming Fishery Management Policy." M.S. Thesis (Ithaca, N.Y. Cornell University) 81 pp.

Funk, John L. 1970. "Warm Water Streams," in Norman G. Benson, ed., A Century of Fisheries in North America (Washington, D.C., American Fisheries Society) pp. 141-152.

Gianessi, Leonard P., Henry M. Peskin and G. Kenneth Young. 1981. "Analysis of National Water Pollution Control Policies. 1.: A National Network Model," Water Resources Research vol. 17 (August) pp. 796-802.

Gianessi, Leonard P. and Henry M. Peskin. 1981. "Analysis of National Water Pollution Control Policies. 2.: Agricultural Sediment Control," Water Resources Research vol. 17 (August) pp. 803-821.

Green, B. L. and H. A. Wadsworth. 1967. "Boaters, Fishermen, Hunters--What Affects Participation and What do they Want?" Agriculture Experiment Station Research Bulletin No. 829. (W. LaFayette, Ind., Purdue University).

Nielsen, Larry. 1980. "Water Quality Criteria and Angler Preference for Important Recreational Fishes." EPA Benefits Project Recreation Working Paper #3. Unpublished. (Washington, D.C., Resources for the Future).

O'Dell, Jay, R. Dintaman, and J. Gabor. 1976. "Fishes of the Potomac Estuary," in William T. Mason and Kevin C. Flynn, eds., The Potomac Estuary: Biological Resources. Proceedings of A Symposium sponsored by the Interstate Commission on the Potomac River Basin and Maryland Department of Natural Resources Power Plant Siting Program, Alexandria, Virginia, June 4-6, 1975 (ICPRB Technical Publication 76-2) pp. 100-104.

Ohio River Valley Sanitation Commission. 1978. Summary of Ohio River Fishery Surveys, 1968-76. Report prepared for the U.S. Environmental Protection Agency, Surveillance and Analysis Division, Philadelphia, Pa. (Springfield, Va., National Technical Information Service, June, PB-288872).

Phillips, C. and S. E. Ferguson. 1977. "Hunting and Fishing Expenditure Values and Participation Preferences in Wyoming, 1975". (Laramie, Wyo., University of Wyoming, Water Resource Research Institute) 184 pp.

Pindyck, Robert S. and Daniel L. Rubinfeld. 1976. Econometric Models and Economic Forecasts (New York, McGraw-Hill).

Rainwater, F. H. 1962. "Stream Composition of the Coterminous United States," in U.S.G.S. Hydrologic Investigation Atlas HA-61 (Washington, D.C., U.S. Department of the Interior, Geological Survey).

Resources for the Future. 1980. Fishable Water Survey (unpublished data).

U.S. Department of Agriculture, Soil Conservation Service. 1973. Soil Survey Laboratory Data and Descriptions for Some Soils of California (June).

U.S. Department of Commerce, Bureau of the Census. 1972. Public Use Samples of Basic Records from the 1970 Census (Washington, D.C.) 203 pp.

U.S. Department of the Interior, Fish and Wildlife Service. 1968. National Survey of Needs for Hatchery Fish. Bureau of Sport Fisheries and Wildlife, Publication No. 63 (Washington, D.C.)

U.S. Environmental Protection Agency. 1974. Development Document for Effluent Limitations Guidelines and New Source Performance Standards for the Steel Making Segment of the Iron and Steel Manufacturing Point Source Category. EPA-44011-74-024-2, (Washington, D.C.) 461 pp.

______. 1975. Cost Estimates for Construction of Publicly-owned Treatment Facilities (Washington, D.C.)

______. 1976. Quality Criteria for Water (Washington, D.C.) 256 pp. (The Red Book).

______, Office of Planning and Evaluation. 1980. National Accomplishments in Pollution Control: 1970-1980. Draft Report (Washington, D.C.).

U.S. Geological Survey. 1972 and 1973. Water Resources Data: Part 2 - Water Quality Records (44 volumes by state, PB 272898 through PB 286880) (Springfield, Va., National Technical Information Service).

Volk, A. A. and V. E. Montgomery. 1973. The Economic Impact of Sport Fishing in South Dakota, 1972, with Notes on Angler Traits (Pierre, S.D., South Dakota Dept. Game, Fish and Parks, Federal Aid to Fish Restoration project E-21-R-7 (study II, job 1)) 60 pp.

Water Research Centre. 1979. "New European Standards for Freshwater Fish," Notes on Water Research, No. 20 (February), (Stevenage, Herts., England, Water Research Centre).

Wisconsin Department of Natural Resources, Bureau of Fish Management. 1978. Wisconsin Lakes, Publication 7-3600(78), (Madison, Wis.).

Chapter 3

THE PROBABILITY OF PARTICIPATION IN RECREATIONAL FISHING

The first choice in the individual's chain of decisions about recreational fishing, as outlined in chapter 1, is whether or not to do some fishing. This decision logically precedes the decisions of what species of fish to seek and how much time to spend per year seeking them. This chapter discusses the modeling of the fishing participation decision using logit and ordinary least squares (OLS) binary choice models.[1]

It should be recognized that the sequential decision model (explained in Cragg, 1971) may be a bit strained and artificial, even though it traditionally has been employed in empirical recreation participation analysis. The model assumes that the decisions on whether or not to participate and how much to participate are independent, not interrelated. Tradition is followed in this chapter and the next because our data come from two unmatchable samples of unequal size, making the multistep procedure necessary.

We hypothesize that the individual's decision about whether to engage in fishing or not is sensitive to the opportunity to fish, represented by the quantity of fishable water at his disposal. A theoretical justification for this use of supply side variables can be found in Deyak and Smith's (1978) adaptation of Becker's (1965) household production model discussed in chapter 1 and appendix A.

It is interesting to note that the more conventional approach to the supply and demand for recreational service flows, which does not employ the household production function approach, still leads one to a reduced-form equation for quantity demanded which depends, among other things, on a quantity measure of the supply of recreational opportunities available (Cicchetti and coauthors, 1972; Cicchetti, Fisher, and Smith, 1973; Cicchetti, 1973). So, both structural models lead to reduced forms with similar arguments. Since we are concerned only with the reduced form here, the choice of a structural model is a matter of taste. Suffice it to say that there is a convincing body of theoretical and applied literature justifying the inclusion of an exogenous variable that explains supply availability in the reduced form.[2]

Our indicator of the quantity of fishable water available to any individual is a crude proxy measured at the state level and is obtained by dividing the total acreage of fishable water in each state reported in Hatchery Fish (U.S. Department of the Interior, 1968) by the 1975 population of each state. This level of aggregation in the supply variable has to be used because the available water quality information does not permit the identification of fishable water at any finer level (county, for example), nor do the individual participation data (USDI, n.d.) provide information on where individuals reside or where they recreate within their states of residence.

Thus, the level of aggregation in our supply variable involves the implicit assumption that fishable freshwater is evenly distributed throughout the state such that all residents in a given state face similar "supply" situations. If one had more detailed information on both the within-state distribution of fishable water and individuals, more representative state-level aggregates could be constructed. On an even more disaggregated level, if every individual could be connected with a measure of water availability particular to him, then a separate participation choice equation could be estimated for every state, and across-state differences in the participation response to water availability could be tested statistically. Neither of these options is possible given the degree of aggregation in the data currently available.

Another problem which must be faced when modeling participation choice equations is that economic theory provides only rough guidelines for specifying the relevant variables influencing participation choices, and often the data set at hand does not include variables that theory suggests could be relevant. For example, it is reasonable to hypothesize that one variable influencing an individual's decision to fish is whether or not his or her parents showed an active interest in such activities, and presumably stimulated a similar interest in their progeny. Measures of such "inherited" interest in recreational fishing are not available in the data set from which we estimate our recreation participation functions: the 1975 National Survey of Hunting, Fishing, and Wildlife Associated Recreation (NSHWFR).[3] This, however, is only one of the problems with NSHFWR, as we explain below.

THE NSHFWR DATA SET

The NSHFWR was conducted in two phases. In the first, a telephone survey of 106,294 households (over 313,000 individuals) was conducted to determine whether or not individuals participated in hunting, fishing, and other wildlife-associated recreational activities. In the second phase, a mail questionnaire was sent to 56,575 individuals nine years of age or older who said in answer to the phone survey that they were hunters and fishermen. Detailed information on their participation patterns, socioeconomic characteristics, and preferences was solicited. This two-stage structure defines our three-stage estimation procedure. We use the telephone survey data in Step I to estimate the probability of a person's being a fisherman, while we use the more detailed information on fishermen only from the mail survey to estimate our Step II and Step III equations, which are not discussed until the next chapter.

Unfortunately, the NSHFWR sample design was done with the production of individual state reports in mind, and the aim was to conduct enough interviews in each state to allow significant statistics to be calculated at the state level for some disaggregated measures. But, presumably for reasons of economy, no attempt was made to interview larger numbers in larger states. The result is a sample in which approximately 2,000 househould phone interviews were conducted in each state, so that smaller states are overrepresented. In addition, residents of nonmetropolitan areas were sampled at rates two to three times their actual importance in each state's population to improve the efficiency of statistics produced from the survey. While the use of such stratified sampling methods is perfectly appropriate given the purpose of the survey -- the production of state and national reports on numbers of sportsmen and related information -- it creates problems in our application. Particularly, the actual recorded observations overrepresent that part of the population living in rural areas and small states--as a general matter the part that other studies would lead us to expect is more likely to produce fishermen (and hunters).

Table 3-1 shows how the NSHFWR data set differs from census data in terms of each state's share of the total population, giving too much weight to observations in sparsely populated states like Maine, New Hampshire, and Wyoming, and too little weight to observations in more populous states like California, Ohio, and New York. The table also shows the disproportionate weight placed on rural residents in the NSHFWR sample, relative to census

Table 3-1. Comparison of U.S. Census State Population Shares and Percent Metropolitan Area Residents with NSHFWR Sample

State	State population as a percent of total U.S. population		Metropolitan state population as a percent of state population	
	Census	NSHFWR	Census	NSHFWR
Alabama	1.71	1.81	61.8	45.4
Arizona	1.04	1.89	74.5	65.0
Arkansas	1.00	2.02	38.4	16.9
California	10.01	2.10	92.8	86.4
Colorado	1.20	1.99	80.6	58.0
Connecticut	1.46	1.94	88.3	52.5
Delaware	0.27	2.01	68.9	54.1
District of Columbia	0.34	0.43	100.0	100.0
Florida	3.91	1.76	86.0	71.0
Georgia	2.33	1.96	56.8	46.7
Idaho	0.38	2.07	16.8	19.1
Illinois	5.29	4.26	81.3	60.5
Indiana	2.51	2.12	67.8	33.3
Iowa	1.35	2.08	37.4	36.0
Kansas	1.08	1.90	45.9	34.9
Kentucky	1.60	1.83	45.5	34.2
Louisiana	1.80	2.25	63.4	46.1
Maine	0.50	1.97	23.3	19.3
Maryland	1.95	1.81	85.1	69.0
Massachusetts	2.75	2.10	86.2	55.3
Michigan	4.30	1.92	81.4	62.5
Minnesota	1.85	1.93	64.4	46.0
Mississippi	1.11	2.09	26.5	19.0
Missouri	2.25	1.97	63.5	29.9
Montana	0.35	1.87	24.4	25.6
Nebraska	0.73	1.90	44.7	25.7
Nevada	0.28	2.07	80.9	61.4
New Hampshire	0.38	2.21	36.3	4.8
New Jersey	3.46	2.02	92.2	81.3
New Mexico	0.54	2.31	33.6	23.3
New York	8.53	2.08	88.5	73.1
North Carolina	2.57	1.84	45.3	25.1
North Dakota	0.30	1.99	22.7	10.7
Ohio	5.07	1.86	79.7	57.3
Oklahoma	1.28	1.93	55.6	57.5
Oregon	1.08	1.90	59.9	60.7
Pennsylvania	5.60	2.05	80.5	64.2
Rhode Island	0.44	1.93	92.3	72.9
South Carolina	1.33	1.99	48.3	36.6
South Dakota	0.32	1.96	14.7	8.5
Tennessee	1.97	1.92	63.1	24.7
Texas	5.78	1.90	79.1	39.3
Utah	0.57	2.26	78.8	67.9
Vermont	0.22	1.92	0	0
Virginia	2.35	2.09	65.6	42.7
Washington	1.68	3.98	71.4	40.2
West Virginia	0.85	1.87	36.4	24.9
Wisconsin	2.17	2.01	60.2	49.0
Wyoming	0.18	1.95	0	0
TOTAL	100.00	100.00	73.0[a]	43.2[a]

Sources:

CENSUS: U.S. Department of Commerce, Bureau of the Census. *Statistical Abstract of the United States, 1978*. Tables 11 and 19.

NSHFWR: Based on 306,087 responses in U.S. Department of the Interior, *1975 National Survey of Hunting, Fishing and Wildlife Associated Recreation* data tape.

[a]Weighted average

data. These disproportions in the raw NSHFWR data influence the proportion of fishermen calculable from each state's sample observations. Fortunately, we can demonstrate this effect because the NSHFWR survey contains individual "weights" which adjust the raw data to reflect the probability of an observation's occurrence had the grand sample been designed in a proportional random rather than a stratified nonrandom manner (see Williams, 1978).[4] A comparison between the raw and weight-adjusted state frequencies of recreational fishing participation appears in table 3-2. We see from this comparison that the overrepresentation of nonmetropolitan area residents in the raw NSHFWR data biases the probability of fishing participation upward, because nonmetropolitan area residents are more inclined to fish than their urban counterparts.

USING THE NSHFWR DATA SET

Whatever the faults of the national telephone survey part of NSHFWR, its size is overwhelming, there being over 313,000 individual interview records from the calls to 106,294 households. Using this many observations to estimate a choice probability model would have been too expensive, so we first set out, as have others (for example, Hay and McConnell, 1979) to draw a selective subsample by choosing every n^{th} observation. This technique, of course, reproduces on a smaller scale the nonproportional characteristics of the larger data set.

Subsample design would not matter much for econometric estimation if the coefficients in the behavioral model were homogeneous throughout the population (Porter, 1973). The expected value of a cross-section regression coefficient estimator is a weighted average of "corresponding" and "noncorresponding" coefficients in the unknown behavioral equations of individual micro units (Zellner, 1962), so aggregation does no damage if all of the micro equations are in fact identical.

The functions we propose to estimate for participation in freshwater recreational fishing are best viewed as representative national functions, where the micro unit of measurement for the key supply side variable, acres of fishable freshwater per capita, is measured at the state level, while the micro unit for all other variables is the individual. Thus, within-and-across state aggregation is implicit in the participation probability equations estimated using a subsample drawn from the NSHFWR

Table 3-2. Proportion of Fishermen of All Ages in Total State Population from Telephone Survey Data

State	Proportion of fishermen	
	Unweighted	Weighted
Alabama	0.373	0.353
Arizona	0.289	0.274
Arkansas	0.378	0.366
California	0.278	0.269
Colorado	0.338	0.324
Connecticut	0.240	0.221
Delaware	0.282	0.281
District of Columbia	0.110	0.105
Florida	0.350	0.325
Georgia	0.325	0.312
Idaho	0.443	0.441
Illinois	0.301	0.267
Indiana	0.330	0.322
Iowa	0.396	0.389
Kansas	0.414	0.408
Kentucky	0.299	0.288
Louisiana	0.383	0.382
Maine	0.314	0.301
Maryland	0.242	0.228
Massachusetts	0.247	0.224
Michigan	0.354	0.334
Minnesota	0.465	0.467
Mississippi	0.333	0.329
Missouri	0.382	0.362
Montana	0.406	0.402
Nebraska	0.341	0.340
Nevada	0.317	0.294
New Hampshire	0.304	0.306
New Jersey	0.213	0.196
New Mexico	0.269	0.270
New York	0.250	0.233
North Carolina	0.307	0.311
North Dakota	0.410	0.405
Ohio	0.276	0.272
Oklahoma	0.422	0.416
Oregon	0.347	0.346
Pennsylvania	0.206	0.194
Rhode Island	0.213	0.195
South Carolina	0.301	0.290
South Dakota	0.352	0.345
Tennessee	0.311	0.299
Texas	0.381	0.369
Utah	0.380	0.383
Vermont	0.343	0.337
Virginia	0.290	0.293
Washington	0.396	0.358
West Virginia	0.245	0.242
Wisconsin	0.392	0.384
Wyoming	0.410	0.408
Average	0.332	0.305

Source: U.S. Department of the Interior. 1975 National Survey of Hunting, Fishing and Wildlife Associated Recreation, data tape.

cross-section.[5] The potential danger is that if, say, the coefficients in rural individuals′ participation functions differ markedly from those of urban individuals, and the former are overrepresented in the sample relative to their representation in the total population, the estimated coefficients of the average aggregate function will be biased accordingly.[6]

ESTIMATING A FISHING PARTICIPATION EQUATION

Before we can develop an equation to predict how changes in the amount of fishable water affect participants in freshwater fishing, we must come to grips with three questions.

> What are the relevant, available explanatory variables for the models?
>
> Does subsample design matter?
>
> What method of subsample selection is to be preferred?

We answer these questions in the following sections, emphasizing the last two.

Choosing Variables for the Prediction Equation

Theoretical considerations (Deyak and Smith, 1978), previous empirical work (Hay and McConnell, 1979; Settle, 1980), and the limitations of the data set at hand led us to the list of potentially important independent variables presented in table 3-3. Note that for compactness the list includes the variables used in this chapter, designated as Stage I, as well as the variables used in the next chapter, which discusses Stages II and III.

The reasons for including the socioeconomic variables referring to age, sex, income, status of the individual in the household, and region of residence are discussed at length in the above sources. The justification for the supply variable, acres of fishable freshwater per capita (INACRE), has been presented in chapter 1. Since the sample also contains saltwater and Great Lakes fishermen, a proxy supply variable called COAST, analogous to INACRE but representing access to nonfreshwater or Great Lakes fishing, has also been included. Finally, we conjectured that however well endowed with fishable water an individual′s state might be, he would be more likely to be a fisherman if surrounding states were amply endowed than if they were not. The variable OUTACRE is intended to capture this potential effect. In addition, one could reasonably argue that whether

Table 3-3. Dependent and Independent Variables Used in Stage I, II, and III Equations

A. DEPENDENT VARIABLES

Name	Description	Source[a]	Stage used[b]
FISH	Binary choice 1 if fisherman; 0 if otherwise	NSHFWR Phone	I
TMAN	Binary choice of doing "some" cold-water game fish angling or not. Equal to: 1, if any days spent cold-water game fish angling 0, otherwise	NSHFWR Mail	II
BPMAN	Binary choice of doing "some" warm-water game fish/panfish angling or not. Equal to: 1, if any days spent warm-water game fish/panfish angling or not. 0, otherwise	NSHFWR Mail	II
CMAN	Binary choice of doing "some" rough fish angling or not. Equal to: 1, if any days spent rough fish angling 0, otherwise	NSHFWR Mail	II
TROUTDA	Days spent per person per season fishing for cold-water game fish in fresh-water (i.e., not including days spent on the Great Lakes or saltwater) for fishermen with TMAN = 1.0	NSHFWR Mail	III
BAPADA	Days spent per person per season fishing for warm-water game fish/panfish in freshwater (i.e., not including days spent on the Great Lakes or saltwater) for fishermen with BPMAN = 1.0	NSHFWR Mail	III
ROUGHDA	Days spent per person per season fishing for rough fish (i.e., not including days spent on the Great Lakes or saltwater) for fishermen with CMAN = 1.0	NSHFWR Mail	III

B. INDEPENDENT VARIABLES

Name	Description	Source	Stage Used
INCH	Respondent's household income, 1975 dollars	NSHFWR Phone, Mail	II, III
AGE	Respondent's age, years	NSHFWR Phone, Mail	I, II, III
AGESQ	AGE squared	NSHFWR Phone, Mail	I, II, III
SEX	Dummy variable for sex: 0 = male; 1 = female	NSHFWR Phone, Mail	I, II, III

Table 3-3 continued

B. INDEPENDENT VARIABLES (cont'd.)

Name	Description	Source	Stage Used
METRO	Dummy variable to indicate if respondent lives in a metropolitan or nonmetropolitan area: 1 = metro resident; 0 = otherwise	NSHFWR Phone, Mail	I, II, III
INCPER	Average income per household member (i.e., household income divided by persons per household)	NSHFWR Phone	I
HEAD	Dummy variable to indicate if respondent is head of household 1 = head of household; 0 = otherwise	NSHFWR Phone	I
PREF	Dummy variable to indicate the intensity of preference for fishing of any sort over other wildlife-associated recreational activities: 1 = fishing is favorite activity; 0 = otherwise	NSHFWR Mail	III
WEST	Dummy variable for residence in Census West Region: 1 = live in West; 0 = otherwise	NSHFWR Phone	I
CENT	Dummy variable for resident in Census Central Region: 1 = live in Central; 0 = otherwise	NSHFWR Phone	I
SOUTH	Dummy variable for resident in Census South Region: 1 = live in South; 0 = otherwise	NSHFWR Phone	I
COAST	Dummy variable to indicate whether the respondent's state has a marine and/or Great Lakes coastline or not: 1 = marine and/or Great Lakes coastline; 0 = otherwise	Maps	I, II, III
INACRE	Total acreage of fishable non-Great Lakes freshwater per capita in respondent's state. A measure of the availability of in-state supply	<u>Hatchery Fish</u>	I
OUTACRE	1, if fishable acres per capita in bordering states exceeds INACRE; 0, otherwise[c]	<u>Hatchery Fish</u>	I
TBAG	Average number of cold-water game fish caught per fishing day by respondents favoring the activity, 11 regional averages	NSHFWR Mail	II, III
BPBAG	Average number of warm-water game fish/panfish caught per fishing day by respondents favoring the activity, 11 regional averages	NSHFWR Mail	II, III

Table 3-3 continued

B. INDEPENDENT VARIABLES (cont'd.)

Name	Description	Source	Stage Used
CBAG	Average number of rough fish caught per fishing day by respondents favoring the activity, 11 regional averages	NSHFWR Mail	II, III
RFFBPSHR	Ratio of fishable freshwater acreage in respondent's state suitable for warm-water game fish/panfish angling to total state fishable acreage, all species	RFF Survey[d]	II
RFFRSHR	Ratio of fishable freshwater acreage in respondent's state suitable only for rough fish angling to total state fishable acreage, all species	RFF Survey[d]	II
RFFTRAC	Fishable freshwater acreage per capita in respondent's state suitable for cold-water game fish angling	RFF Survey	III
RFFBPAC	Fishable freshwater acreage per capita in respondent's state suitable for warm-water game fish/pan fish angling.	RFF Survey	III
RFFRAC	Fishable freshwater acreage per capita in respondent's state suitable only for rough fish angling.	RFF Survey	III

[a] Sources:

NSHFWR Phone: U.S. Fish and Wildlife Service. 1975 National Survey of Hunting, Fishing and Wildlife Associated Recreation (Data Tape, Telephone Interviews).

NSHFWR Mail: U.S. Department of the Interior, Fish and Wildlife Service. 1975 National Survey of Hunting, Fishing and Wildlife Associated Recreation. (Data tape, Mail Questionnaires)

Hatchery Fish: U.S. Fish and Wildlife Service. 1968. National Survey of Needs for Hatchery Fish, 1965. This source estimates, state-by-state, the available acres of fishable freshwater as discussed in Chapter 2.

RFF Survey: Resources for the Future. 1980. RFF Fishable Water Survey (Unpublished Data).

RFFWQNM: Leonard P. Gianessi, Henry M. Peskin, and G. K. Young. "Analysis of National Water Pollution Control Policies: 1. A National Network Model," Water Resources Research vol. 17, no. 4. (August 1981) pp. 796-801.

Table 3-3 continued

Notes to Table 3-3 (cont'd.)

[b] The stages of the individual's decisions which are modeled are:

Stage I. Estimation of an equation for predicting the probability of being a fisherman

Stage II. Estimation of equations for predicting the probability of doing some:

a. Cold-water game fishing (trout)
b. Warm-water game or pan fishing
c. Rough fishing (carp, catfish)

given that one is a fisherman.

Stage III. Estimation of equations for predicting the average days per angler spent fishing for:

a. Cold-water game fish
b. Warm-water game fish or pan fish
c. Rough fish

OUTACRE was defined as follows for state j: Let

$$OA = \frac{\sum_{i=1}^{n} \text{fishable acreage in the n states bordering j}}{\text{Population of j} + \sum_{i=1}^{n} \text{population of the n bordering states}}$$

If OA > INACRE then OUTACRE = 1
IF OA ≤ INACRE then OUTACRE = 0

[d] The survey was supplemented where necessary with information from Hatchery Fish and the RFFWQNM.

a person fishes or not depends in part on the availability of recreational substitutes. Lack of data on availability of other recreational resources for activities such as boating, swimming, golf, or tennis prevented their inclusion, however. Our model effectively assumes the absence of cross-price effects in the demands for final recreational service flows of various sorts and the absence of joint production on the supply side.

Testing the Impact of Subsample Design

A second critical question is, does subsample design make a significant difference in the estimation of participation probability and, particularly, in the values for the coefficients on key policy-related variables?

Two subsamples of roughly equivalent size were drawn from the NSHFWR telephone survey grand sample to explore this question. (For a more detailed discussion, see Vaughan and Russell, 1981.) The first involved a systematic draw of every 45th record following Hay and McConnell's (1979) procedure, which resulted in a selective, nonproportional subsample of 5,576 observations reflecting the peculiarities of the grand sample. We also designed a sampling procedure that produced a subsample of 5,364 observations more nearly matching the actual proportions of states, and of metropolitan-area residents within states, in the total U.S. population. To guard against the possibility that the raw data in NSHFWR were arranged according to some unstated pattern, this sample was drawn randomly rather than systematically.[7]

The question we want to test is whether the estimated probability-of-participation functions for fishing, estimated by both OLS and logit, are significantly different between the selective and proportional samples. The null hypothesis is that they are not. One way to answer this question is to produce restricted and unrestricted estimates from a pooled data set composed of the selective and proportional samples, where the restriction of equal coefficients across samples is relaxed using an interactive "group" dummy variable (Johnston, 1972, pp. 192-207; Gujarati, 1970).[8]

Below, we report an F-statistic for the test of the significance of the difference between the entire selective and proportional OLS equations (Chow, 1960; Fisher, 1970) and a log-likelihood chi-square statistic (Nerlove and Press, 1973) for the same test across logit equations.[9]

Table 3-4. Test Results: Overall Fishing Equations

I. OLS Model	Sum of Squared Residuals[a]		F
	Restricted	Unrestricted	
	2165.06	2160.63	3.73[b]
II. Logit Model	Log of Likelihood Function[a]		χ^2
	Restricted	Unrestricted	
	-6340.65	-6329.71	21.89[b]

Notes:

[a] T = Total number of observations in a pooled sample composed of the proportional and the selective samples = 10940.

k = Number of explanatory variables in the restricted equation = 6.

m = Number of variables tested with the dummy = 6.

[b] Significant at 1 percent level.

These results show that, taken overall, the participation equation for fishing is significantly different between the proportional and selective samples based on the F(OLS) and χ^2 (logit) tests of homogeneity. We infer that the microunit fishing probability functions differ across individuals. This may be because fishing is more common among those individuals with a nonmetropolitan background and because of differences across states in the relevant parameters of the microfunctions. Second, and at least as important, the magnitude of the coefficient attached to the key policy variable, INACRE, in the probability-of-participation equation for fishing is severely influenced by sample design. The proportional sample results (not reported here) suggest that a one-acre increase in fishable water per capita would lead to an 18 percentage point increase in fishing frequency, while the distorted selective sample produces an increase in fishing frequency of only 7 percentage points -- a 60 percent understatement of the impact of additional acreage on participation.

In the only other similar study we have seen, Davidson, Adams, and Seneca (1966), using a different data base and set of explanatory variables, estimated an OLS probability-of-fishing equation which produced a coefficient of 0.23 on the acreage of fishable water per capita. This supplementary evidence suggests that an unbiased national average coefficient is more properly estimated from the proportional sample. A better solution, if disaggregated supply side data are ever developed, would be to estimate such functions on a state-by-state basis.

FINAL LOGIT ESTIMATION

To estimate a logit equation for use in prediction, we drew a proportional sample of 7,760 observations in the same manner used to construct the 5,364 observation proportional sample discussed previously. The sample size was increased by nearly 50 percent in order to obtain more efficient coefficient estimates. The means of the variables in our large sample match fairly well with known census values, as shown in table 3-5.

Three alternative specifications of the final equation itself appear in table 3-6. The first is the full model employing all of the variables listed in table 3-3. The second (Reduced Model I) posits the null hypothesis that the coefficient attached to OUTACRE is zero,[10] and the third posits the null hypothesis that the coefficients attached to OUTACRE, COAST, HEAD, and INCPER are all zero. Chi-square tests of the zero restrictions do not allow rejection of the null restriction on the OUTACRE coefficient of Reduced Model 1 at the 2.5 percent level. It is not surprising that the OUTACRE variable measuring the relative attractiveness of bordering states' supply is not significant when we consider the fact that over 90 percent of all fishing activity takes place within the individual's state of residence. (See footnote 5.)

The set of restrictions imposed in Reduced Model II vis-a-vis the full model cannot be rejected at the 2.5 percent level.[11] In our view, there is little reason not to choose Reduced Model II as a preferable simplification of the specification of the full model. We found no evidence of strong near-dependencies (collinearity) among the explanatory variables in the model based on singular value decomposition (SVD) of the matrix of observations on the independent variables (Belsley, Kuh, and Welsch, 1980).[12] Therefore, the variable deletion involved to arrive at Reduced Model II probably does not severely bias the estimates. However, because a strong case can be made against pretesting in regression analysis (ad hoc interpretive searching over the parameter space), we use Reduced Model I as the preferred specification in the benefit evaluations of chapter 6. Using either of the other models would not change the benefit results significantly.

Table 3-5. Means of Variables

Variable	Description	Means: Proportional sample	Means: Census
I. DEPENDENT			
FISH	1 if fisherman, 0 otherwise	0.303	N.A.
II. INDEPENDENT			
AGE	Age of respondent, years	31.38	32.87
AGESQ	Age squared[a]	984.70	1080.47
SEX	1 if female, 0 otherwise	0.507	0.51
METRO	Fraction metropolitan area residents	0.715	0.730
INACRE	Acres of fishable freshwater/ state population	0.148	N.A.
WEST	1 if resident of census West Region, 0 otherwise	0.177	0.178
CENT	1 if resident of Census Central Region, 0 otherwise	0.284	0.271
SOUTH	1 if resident of Census South Region, 0 otherwise	0.323	0.320
HEAD	1 if household head, 0 otherwise	0.341	0.336
COAST	1 if marine or Great Lakes coastline, 0 otherwise	0.811	N.A.
OUTACRE	1 if fishable acres per capita in bordering states exceeds INACRE, 0 otherwise	0.571	N.A.
INCPER	Income per person, 1975 dollars	4571.19	4814

Sources for Census Data:

AGE: Current Population Reports, Population Characteristics Series P-20, No. 334, January 1979. The mean value for the open-ended class "65 and over" was taken to be 75 years on the breakdown of the 65 to 85 year range reported in Current Population Reports, Special Studies Series P-23, No. 59, May 1976.

All other variables: U.S. Department of Commerce, Bureau of the Census, Statistical Abstract of the United States, 1978.

[a]This is the square of the mean age, though the variables actually used were, of course, the squared age of each respondent.

Table 3-6. Final Probability of Fishing: Proportional Sample Participation Equation ("t" statistics in parentheses)

	Full Model		Reduced Model I		Reduced Model II	
Variable	Coefficient	Partial Derivative[b]	Coefficient	Partial Derivative[b]	Coefficient	Partial Derivative[b]
INTERCEPT	-1.07954 (8.23)*	...	-1.07754 (8.60)*	...	-1.16515 (10.79)*	...
AGE	0.05515 (9.99)*	0.01113	0.05515 (9.99)*	0.01110	0.05068 (10.25)*	0.01022
AGESQ	-0.00078 (11.15)*	-0.00016	-0.00078 (11.15)*	0.00016	-0.00075 (11.16)*	-0.00015
SEX	-1.25169 (19.80)*	-0.25272	-1.25168 (19.80)*	-0.25197	-1.17824 (21.98)*	-0.023753
METRO	-0.30602 (5.16)*	-0.06179	-0.30606 (5.16))*	-0.06161	-0.31538 (5.41)*	-0.06358
WEST	0.43239 (4.86)*	0.08730	0.43279 (4.89)*	0.08712	0.46236 (5.33)*	0.09321
CENT	0.52995 (6.66)*	0.10700	0.53004 (6.67)*	0.10670	0.55247 (7.06)*	0.11138
SOUTH	0.40790 (5.20)*	0.08236	0.40777 (5.20)*	0.08209	0.42297 (5.46)*	0.08527
INACRE	0.89417 (6.11)*	0.18054	0.89064 (6.91)*	0.17929	0.92382 (7.25)*	0.18624
INCPER	0.0000079 (1.14)	0.0000016	0.0000079 (1.14)	0.0000016	...	...
HEAD	-0.17347 (2.29)*	-0.03502	-0.17346 (2.29)*	-0.03492	...	...
COAST	-0.13345 (1.96)*	-0.02694	-0.13312 (1.97)*	-0.02680	...	...
OUTACRE	0.00313 (0)	0.00063	...	...	...	...
Chi-square test	814.08*		814.08*		804.00*	
Likelihood ratio index	0.086		0.086		0.084	
Probability of fishing evaluated at means[a]	0.281		0.279		0.280	

Note: An asterisk (*) indicates significance at the 5 percent level.

[a] $P_{FISH} = \dfrac{1}{1+e^{-(INTERCEPT + \sum_{i=1}^{n} \hat{\beta}_i \bar{X}_i)}}$. Actual sample frequency (table 3-5) is 0.303.

[b] The partial derivative of the logit model is $\partial p/\partial X_i = p(1-p)\beta_i$ where p is the probability of participation and β_i is the estimated parameter of the ith independent variable, X_i, in the logit model.

Closer inspection of Reduced Model I shows that the overall equation is significant and that eleven of the twelve parameter estimates are significant at the 5 percent level. The likelihood ratio index, an analog to the coefficient of determination (R^2) in linear modeling, (Domenich and McFadden, 1975, pp. 123-125) does not inspire overwhelming confidence in the forecasting ability of the overall equation, but we do not intend to use the equation for forecasting with simultaneous changes in all of the explanatory variables. Rather, our interest focuses on the coefficient attached to the supply variable, INACRE, since in our framework water pollution control policy works through the availability of fishable freshwater, and our benefit analysis holds all other variables at their base (mean) values. It is important that the parameter estimate on INACRE is statistically significant with the expected positive sign.

The impact of the INACRE supply variable (acres of fishable freshwater available at the state level per state resident) is displayed in figure 3-1

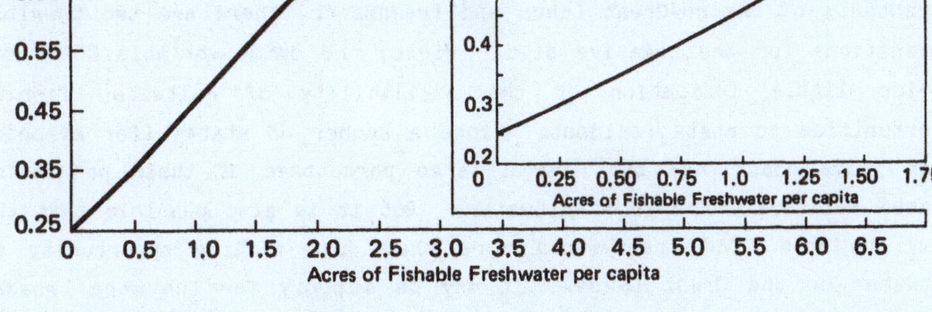

Figure 3-1. Logistic curve. Probability of fishing as a function of available fishable freshwater per capita. Reduced Model II.

holding all other variables constant at their means. This function is almost linear over the relevant range of observations on supply per capita, suggesting that at least for this data set, OLS estimation of the dichotomous dependent variable model might have been as accurate as the more expensive logit approach actually undertaken, confirming the remarks in Smith and Munley (1978). (See also chapter 4, footnote 6).

While we are less interested in the coefficients on the remaining independent variables, some discussion of them is warranted. Our finding that income per capita (INCPER) is not a useful explanatory variable is an unusual one, for income is ubiquitous in predictive equations for all sorts of human activity choices.[13] There are two possible explanations for the income result. Most important, when aggregating across all types of fishing, as is done for this equation, income may not in fact be an important determinant of participation. Because the water body involved can be close to home and the equipment the most rudimentary cane pole, while the activity still qualifies as "fishing," low income need not imply much of a limitation on the ability to participate. At the same time, low income may be accompanied by a large amount of leisure, so that the observed income effect after filtration through the household production function would be close to zero.[14] A second possibility is that because the income measure was constructed by dividing household income by the number of persons in the household, and is hence only a proxy for actual per capita income, there may be an error-in-variables problem.[15]

One other unexpected result appears in the probability-of-fishing equation apart from the insignificance of the income variable, and that is the negative (and significant) coefficient attached to COAST. Although one would expect a higher probability of fishing to be associated with the twin attractants of marine-Great Lakes and freshwater, there are two possible explanations for the negative sign. First, the dummy variable COAST may provide little indication of the availability of saltwater fishing opportunities to state residents, since a number of states (for example, Texas) have coastlines but have a large percentage of their population located relatively far from saltwater. But it is also possible that the dummy variable COAST represents more than just fishing opportunity in saltwater or the Great Lakes. It may be a proxy for the more broadly defined opportunity to engage in activities such as surfing, ocean sailing, and just going to the beach, which are unique to saltwater and compete with freshwater activities (including fishing) for the recreationist's time and attention. All the other variables in the equations have signs one would anticipate a priori.

In the next chapter we take up the more complicated question of how the individual, once having decided to be a fisherman, allocates his effort across the various species classes he can fish for.

NOTES

1. It is by now well known that models involving prediction equations for probabilities, where the data are in the form of individual observations and the individuals necessarily did or did not do (chose or did not choose) the activity (alternative) of interest, are in principle best estimated using techniques other than ordinary least squares applied to equations with 0/1 dependent variables. Logit is a technique designed for such estimation problems, and the interested reader can find it described in any of a number of econometric texts (for example, Hanushek and Jackson, 1977, chapter 7). The application of this technique is quite straightforward in the case of a single yes/no, (fish/do not fish) choice. When the choice is among a larger set of alternatives, however, where it may or may not be possible to structure the alternatives as mutually exclusive and exhaustive, and where there may or may not be problems of simultaneous determination of the choices, both theoretical and practical computational problems arise. In this chapter we sidestep these issues by concentrating entirely on the fish/do not fish choice.
   Our initial logit estimation was done using LOGIT, a program written and generously supplied to us by Joseph R. Antos of the U.S. Department of Labor. The remainder was performed using the LOGIST procedure in SAS.

2. See, for example, Deyak and Smith (1978); Miller and Hay (1981); Smith and Munley (1978); Hay and McConnell (1979).

3. Whether or not the omission of this variable would produce biased coefficients on the water availability variables in our participation choice model depends in essence on whether this "inheritance" influence is independent in the long-run, cross-sectional context. If we think of the passing of the "fishing trait" from parent to child as a Markovian process with transition probabilities depending, other things being equal, only on water availability, in the long run the probability that any randomly drawn individual in state i is a fisherman will depend only on water availability in state i.

4. The national survey report summarizing the NSHFWR results comments on page 6 that:

   > In the telephone phase, the weighting model used Census-derived estimates for age, sex and location of residence (metropolitan or non-metropolitan areas). In addition, the results were adjusted by the number of different telephone numbers the sample household had (i.e., the number of chances it had of entering the sample). Thus, each respondent in each state received a weight which took into account his age, sex, place of residence and number of chances (number of telephone numbers) he had of coming into the sample. All estimates were projected to the number of persons in each state for each of the weighting cells.

5. One might argue that a measure which goes beyond state boundaries would be more appropriate than one confined by them. However, calculations based on NSHFWR mail survey data show that 90 percent of the total days of recreational fishing are spent in-state.

6. A contrast between regional and state-by-state recreation demand model coefficients appears in Ziemer and Musser (1979).

7. Each observation in NSHFWR was partitioned by state and metro and nonmetro status to obtain a cell count accurately reflecting the characteristics of NSHFWR. Then we selected a targeted subsample size and, using the U.S. Census proportions, calculated the number of subsample observations within each state which would yield a proportional subsample. For each we then calculated the ratio of the desired number of metropolitan and nonmetropolitan observations to the state totals actually in NSHFWR. By generating 98 separate sets of uniform random numbers (49 states x 2 residency status), each distributed uniformly on the 0-1 interval and including only those observations with a random number less than or equal to the ratio of our desired to the NSHFWR total case representation, we produced our proportional sample with randomized case selection.

8. The dummy variable, D, is set equal to 1.0 for the proportional sample and 0 for the selective sample.

9. The hypothesis that there are no differences between the two samples in the OLS case is based on sums of squared residuals from two separate regressions; one with the dummy variable excluded, to impose the restriction of the null hypothesis (SSR_H), and one with the dummy variable included both additively and interactively (SSR_A):

$$F(m, T\text{-}k\text{-}m) = \frac{\left[SSR_H - SSR_A\right]/m}{SSR_A/(T\text{-}k\text{-}m)}$$

where

T = Number of observations in the pooled sample composed of the proportional and selective samples

k = Number of explanatory variables in the restricted equation

m = Number of variables tested with the dummy

For the logit equation the log-likelihood ratio test of the same null hypothesis as above is:

$$\chi^2(m) \approx -2\left[L(\hat{\theta}_H) - L(\hat{\theta}_A)\right];$$

where:

$L(\hat{\theta}_H)$ = logarithm of the likelihood function at its maximum with the null restriction.

$L(\hat{\theta}_A)$ = logarithm of the likelihood function at its maximum without the null restriction (i.e., with dummy variable included).

10. Our practical reason for imposing this restriction is simply to see if the computational burden of evaluation of the impacts of water pollution control policies on acreage and hence participation can be eased by ignoring OUTACRE.

11. The calculated value of $\chi^2(4df)$ or $-2[(L\hat{\theta}_H) - L(\hat{\theta}_A)]$ is 10.06, and the tabled value of chi-square with 4 degrees of freedom is 11.14 at the 2.5 percent significance level, so for a one-tailed test of the restrictions of Reduced Model II versus the Full Model, we would not reject the null hypothesis.

12. The logit results in table 3-6 show that the parameter estimates are very stable over different regimes of variable specification. This relative insensitivity of the parameter estimates suggests the absence of strong collinearity among the explanatory variables. This observation was confirmed by singular-value decomposition (SVD) of the scaled X matrix (see Belsley, Kuh, and Welsch, 1980).

Using a random subsample of 300 observations from the 7,760 observation sample, we found the largest condition index (ratio of the largest to smallest singular value) to be 24.23, and the next largest to be 12.74. According to Belsley and coauthors, weak dependencies are associated with condition indices between 5 and 10, and moderate to strong relations are associated with condition indices of 30 to 100. Since there are roughly as many near dependencies among the columns of X as there are high condition indices, we have at most one moderate dependency in our data. Examination of the variance decomposition proportions suggested the variable AGE to be involved, which was confirmed by forming the auxiliary regression between age and the rest of the independent variables. The R^2 in this auxiliary regression was 0.9756. This result is not unexpected because a second order term for AGE was in the initial regression.

13. In their fishing probability equation, estimated using OLS and a different data set than that employed here, Davidson, Adams, and Seneca (1966) found income to be significant at the 1 percent level, and COAST at the 10 percent level.

14. This conclusion seems to agree with Settle (1980) where the socioeconomic group variable is not reported as significant in the fishing participation model.

15. Our variable definition follows the procedure used by Hay and McConnell (1979). But, in their equations for the probability of engaging in wildlife watching and wildlife photography, the income variable was significant at the 5 percent level.

REFERENCES

Becker, G. S. 1965. "A Theory of the Allocation of Time," Economic Journal vol. 75 (September) pp. 493-517.

Belsley, David A., Edwin Kuh, and Roy E. Welsch. 1980. Regression Diagnostics: Identifying Influential Data and Sources of Collinearity (New York, Wiley).

Chow, Gregory C. 1960. "Tests of Equality Between Sets of Coefficients in Two Linear Regressions," Econometrica vol. 28 (July) pp. 591-605.

Cicchetti, Charles J., V. Kerry Smith, Jack L. Knetsch, and R. A. Patton. 1972. "Recreation Benefit Estimation and Forecasting: Implications of the Identification Problem," Water Resources Research vol. 8 (August) pp. 840-850.

________, Anthony C. Fisher, and V. Kerry Smith. 1973. "Economic Models and Planning Outdoor Recreation," Operations Research vol. 21, no. 5 (September-October) pp. 1104-1113.

________. 1973. Forecasting Recreation in the United States (Lexington, Mass., Lexington Books).

Cragg, John G. 1971. "Some Statistical Models for Limited Dependent Variables with Application to the Demand for Durable Goods," Econometrica vol. 39 (September) pp. 829-844.

Davidson, Paul, F. Gerard Adams, and Joseph Seneca. 1966. "The Social Value of Water Recreational Facilities Resulting from an Improvement in Water Quality: The Delaware Estuary," in Allen V. Kneese and Stephen C. Smith, eds., Water Research (Baltimore, Md., Johns Hopkins University Press for Resources for the Future).

Deyak, Timothy A. and V. Kerry Smith. 1978. "Congestion and Participation in Outdoor Recreation: A Household Production Function Approach," Journal of Environmental Economics and Management vol. 5, no. 1 (March) pp. 63-80.

Domenich, Thomas A. and Daniel McFadden. 1975. Urban Travel Demand: A Behavioral Analysis (Amsterdam, North-Holland).

Fisher, Franklin M. 1970. "Tests of Equality Between Sets of Coefficients in Linear Relations: An Expository Note," Econometrica vol. 38 (March) pp. 361-366.

Gujarati, Damodar. 1970. "Use of Dummy Variables in Testing for Equality Between Sets of Coefficients in Linear Regressions: A Generalization." The American Statistician vol. 24 (December) pp. 18-21.

Hanushek, Eric A. and John E. Jackson. 1977. Statistical Methods for the Social Sciences (New York, Academic Press).

Hay, Michael J. and Kenneth E. McConnell. 1979. "An Analysis of Participation in Nonconsumptive Wildlife Recreation," Land Economics vol. 55 (November) pp. 460-471.

Johnston, J. 1972. Econometric Methods 3rd ed. (New York, McGraw-Hill).

Miller, Jon R. and Michael J. Hay. 1981. "Determinants of Hunter Participation: Duck Hunting in the Mississippi Flyway," American Journal of Agricultural Economics vol. 63, no. 4 (November) pp. 667-684.

Nerlove, M. and S. J. Press. 1973. Univariate and Multivariate Log-Linear and Logistic Models (Santa Monica, Cal., Rand Corporation).

Porter, Richard D. 1973. "On the Use of Survey Sample Weights in the Linear Model," Annals of Economic and Social Measurement vol. 2 (April) pp. 141-158.

Settle, J. G. 1980. "Relating Participation in Recreational Activities to Social Characteristics," Journal of Environmental Management vol. 10, pp. 138-147.

Smith, V. Kerry and Vincent G. Munley. 1978. "The Relative Performance of Various Estimates of Recreation Participation Equations," Journal of Leisure Research vol. 10, no. 3, pp. 167-176.

U.S. Department of the Interior, Fish and Wildlife Service. 1968. National Survey of Needs for Hatchery Fish. Bureau of Sport Fisheries and Wildlife, Publication No. 63 (Washington, D.C.).

U.S. Department of the Interior. n.d. 1975 National Survey of Hunting, Fishing, and Wildlife-Associated Recreation (NSHFWR) Data tapes and unpublished mimeo, state and national reports (Washington, D.C.).

Vaughan, William J. and Clifford S. Russell. 1981. "An Analysis of Participation in Nonconsumptive Wildlife Recreation: Comment," Land Economics vol. 57 (May) pp. 279-287.

Williams, Bill. 1978. A Sampler on Sampling (New York, Wiley).

Zellner, Arnold. 1962. "Estimation of Cross-Section Relations: Analysis of a Common Specification Error," Metro-economica vol. 14 (April-August-September) pp. 111-117.

Ziemer, Rod F. and Wesley N. Musser. 1979. "Population-Specific Recreation Demand Models and the Consequences of Pooling Sample Data," Western Journal of Agricultural Economics vol. 4 (July) pp. 121-129.

Chapter 4

PARTICIPATION CHOICES AND ANGLING INTENSITY

The individual who has decided to be an angler faces two further decisions in any year: what type of fish to seek and how many days to spend seeking them.[1] On the basis of the same line of argument set out in chapter 1 and pursued empirically for the fish/do not fish decision in chapter 3, we expect to find both these decisions sensitive to the availability of fishable water bodies. Here, however, the more finely specified activities demand more finely specified measures of water availability and we argue that the appropriate measures are different for the two choices.

In this chapter we first develop a model of choice among types of fishing and estimate prediction equations based on that model for use in the benefit calculation exercise in chapter 6. Along the way we discuss the closely related problems posed by the nature of our data sample and the choice of an estimation method. The second major part of the chapter involves a similar chain of discussion and empirical work aimed at the estimation of prediction equations capturing the angler's choice of intensity of participation, that is, of the number of days per year to be devoted to each activity. Here choice of estimation technique is less of a problem.

MODELS OF FISHING CHOICE BY TYPE OF SPECIES SOUGHT

A recreational fisherman may choose among specific activities, both those broadly defined by the distinction between salt and freshwater and, within the latter category, those more finely differentiated by fish type or individual species sought. Thus, at one extreme, a person might specialize in catching brown trout, disdaining even other cold-water game fish, while at the other extreme, a person might take a catholic view of the opportunities available in freshwater, fishing for trout on one day, bass on another, and carp on a third.[2] (Indeed, it is often possible to combine several types of fishing on a single day or even at a single moment.)

A preliminary intuitive picture of the range of angler choices made and their relation to measures of water availability by species supportable is provided in table 4-1. The table includes a selection of eleven states along with the total United States (the lower 48 states plus the District of Columbia.) As a rough rule, and quite as one would expect, the northern tier of states is the one with low enough summer water temperatures to support natural trout fisheries. The southern states have waters dominated by bass or rough fish, depending on natural turbidity and the extent of manmade pollution.

In the matter of pollution, the states vary largely because they have seen very different amounts of industrial development over the past hundred and fifty years, ranging from the chemical heartland of New Jersey (and to a lesser extent Pennsylvania, Illinois, and Alabama) to the more pristine Utah and Montana. The giants, California and Texas, have heavily industrialized regions but also large areas of farming and even wilderness.

All these factors are reflected in the numbers found in table 4-1. The first section shows fishable freshwater acreage--in total, and as a proportion of total freshwater acreage. Further, the table includes an estimate of how that acreage may be divided among the major categories (trout, bass, or rough fish water.) Notice that most of the state fisheries, and the United States overall, are dominated by bass water. Of the eleven states included here, only Maine, Montana, and Utah have more trout than bass water. On the other hand, very little of the total fishable freshwater is suitable only for rough fish. The big surprise in this regard is Montana, with a third of its water classified by state officials as rough fishing. This seems to be because of the importance of the main stem of the Missouri River, which includes a very large reservoir, Fort Peck, created by the U.S. Bureau of Reclamation.

The second section of the table deals with population, in total and in relation to available fishable freshwater. Notice the very large range in this indicator of potential pressure: from about 1/2 person per acre in Maine to almost 100 in Pennsylvania (a factor of 200 difference). But when we look at licensed fishermen per acre, we see a somewhat smaller range (from 0.15 per acre to 8.1 per acre for a factor of about 50).

Table 4-1. Measures of Water Availability, Population, and Sportfishing Pressure for Selected States and the Lower 48 States Combined

	Alabama	Calif.	Illinois	Kentucky	Maine	Montana	Nebraska	New Jersey	Penn.	Texas	Utah	U.S. Totals
Water												
Freshwater fishable area (1,000 acres)[a]	574	863	371	526	1638	1015	176	119	122	1469	372	30,615
Fishable area as proportion of total freshwater area[b]	0.99	0.85	0.98	0.79	0.90	0.996	0.95	0.97	0.79	0.98	0.97	0.93
Proportion of fishable acres classified as[c]												
Trout water	0.01	0.25	0.01	0.02	0.80	0.54	0.05	0.16	0.18	0	0.58	0.20
Bass water	0.95	0.75	0.84	0.77	0.20	0.13	0.35	0.76	0.78	0.80	0.36	0.68
Rough fish water	0.04	0	0.15	0.21	0	0.33	0.60	0.08	0.04	0.20	0.06	0.11
Population												
1975 Population (1,000)[d]	3,614	21,185	11,145	3,396	1,059	748	1,546	7,316	11,827	12,237	1,206	211,904
Persons per freshwater fishable acre	6.3	24.5	30.0	6.4	0.6	0.74	8.8	61.5	96.6	8.3	3.2	6.9
Licensed fishermen per fishable acre[e]	1.0	6.2	2.3	1.3	0.15	0.73	1.5	2.3	8.1	1.0	0.90	1.1
Fishing												
Proportion of population claiming to do some 1976 fishing[f]	0.37	0.28	0.30	0.30	0.31	0.41	0.34	0.21	0.21	0.38	0.38	0.33
Proportion of population 1977 holding resident licenses[e]	0.14	0.25	0.07	0.15	0.15	0.58	0.17	0.04	0.08	0.12	0.27	0.14
Number of nonresident[e] licenses per resident lic.	0.14	0.01	0.05	0.30	0.49	0.73	0.12	0.03	0.06	0.02	0.55	0.17
Proportion of fishermen doing:[f]												
Some trout fishing	0.14	0.72	0.20	0.15	0.86	0.97	0.30	0.35	0.73	0.14	0.98	0.37
Some bass fishing	0.87	0.44	0.86	0.94	0.42	0.18	0.93	0.36	0.65	0.73	0.20	0.68
Some rough fishing	0.49	0.30	0.43	0.53	0.13	0.07	0.62	0.16	0.32	0.44	0.15	0.35
Only saltwater	0.06	0.11	0.04	0	0.08	0.01	0.01	0.49	0.07	0.16	0.01	0.12

[a] U.S. Department of Interior, Fish and Wildlife Service. 1968. National Survey of Needs for Hatchery Fish.

[b] RFF Survey of Fishable Freshwater. Questionnaires sent to state officials, 1980.

[c] As in b. Supplemented by calculations from the RFF pollution source inventory and water quality network model.

[d] U.S. Bureau of the Census.

[e] U.S. Department of Interior, Fish and Wildlife Service. 1978. (News Release) "1977 Hunting and Fishing License Sales Set Records," Washington, D.C.

[f] U.S. Department of Interior, Fish and Wildlife Service. n.d. 1975 National Survey of Hunting, Fishing, and Wildlife Associated Recreation, (Summary documents and data tapes). Washington, D.C.

The third major section of the table pursues the matter of fishing participation in more detail. It shows both the proportion of people who claim to have fished in 1976 and the proportion who bought fishing licenses. The differences in these numbers reflect several things: the fact that "fishing" includes saltwater activity, for which no licenses are required; differences in state enforcement efforts; and probably a tendency to claim participation, even when not justified, in what is thought to be a good, wholesome activity, with perhaps just a hint of machismo about it.[3]

The fishing section also shows the relative importance of nonresident (licensed) fishermen in total fishing pressure. This varies in a roughly inverse way to the potential pressure from the resident population and is essentially an indicator of the attractiveness of the state to fishermen/tourists. (But only roughly. California is apparently not a mecca for out-of-state fishermen though it is for nearly every other kind of tourist.)

Finally, indicators of the importance of trout, bass, and rough fishing to the residents of the several states and the United States overall are included at the bottom of the table. These numbers do not sum to one. They represent, for each of the freshwater fishing categories, the probability that a randomly chosen fisherman would, in 1976, have done <u>some</u> trout fishing (perhaps along with bass, rough, and saltwater fishing), <u>some</u> bass fishing, and <u>some</u> rough fishing. Somewhat asymmetrically the probability that a fisherman would do <u>only</u> saltwater fishing appears below the freshwater categories. Again, the numbers are not terribly surprising if we accept the link between availability and personal choices: in states dominated by trout water, the average fishermen is very likely to do some trout fishing. The same is true for other categories.

Our view of the benefits generated by pollution controls includes as an important component an hypothesized upgrading of the type of fishing available, and an attendant upgrading of activity values (through a higher value assigned to days of fishing for the more sensitive game species introduced or naturally colonizing newly clean water). Therefore it is important to be able to model this second level of choice and especially to obtain quantitative relations between water availability and the choice. As already discussed, our sources of data on base water quality, on participation (and, as we shall discuss in chapter 6, on valuation) are sufficient to support a breakdown of freshwater activity choices no finer than the following:

cold-water game (trout or T) fishing

warm-water game (bass/panfish or BP) fishing

rough (carp/catfish or R) fishing

Our participation data also include saltwater and Great Lakes (S) fishing and in our initial discussion we have to take this into account because of mixed activity possibilities. Later we shall be able to drop this category from explicit consideration.[4]

Given this set of four possible distinct activities, we can define fifteen mutually exclusive categories of angler, one for each possible mix of categories from purist to completely eclectic. These categories appear in the first column of table 4-2, labelled by the letter symbols introduced above.

As a practical matter, however, the data available for estimating the predictive choice equations, which we discuss further below, do not contain enough representatives within some of the mutually exclusive categories to give us variation over the 48 states, the level of our water availability disaggregation. A way around this problem is to define four <u>non</u>mutually exclusive categories of fishing activity based on "any" rather than "only." That is, a person is included in the nonmutually exclusive trout fishing category if he or she does any trout fishing, in the bass/pan category if he does any bass or panfish fishing, and so forth. Then a purist appears in only one category, but the more eclectic a person's tastes the more categories he can be placed in. The formal relationship between the two approaches to categorization is simple. A person is categorized in the ith inclusive (or "some of type i") category if he is a member of one of the eight mutually exclusive categories including some activity type i. In table 4-2, we show how this folding together works by checking the mutually exclusive categories combined into each "some-of" category.

In discussing the results, we show how these two systems are related in estimation. Consider for now two alternative structural models of consumer decision making: a simultaneous and a sequential system. Both models lead to the same reduced-form specification of predictive equations for recreational fishing choice (though we only show equations for the four "some of" categories).

A first hypothesis is that the choices of whether or not to do some fishing in each of the four categories are simultaneously determined. We can reexpress the choice probabilities as the logarithm of the odds that each choice will be made, in order to write a simultaneous system for

Table 4-2. Categories of Fishing Activity: Mutually Exclusive and Nonmutually Exclusive

Mutually exclusive category	Combining mutually exclusive categories			
	Some trout fishing	Some bass/ pan fishing	Some rough fishing	Some salt-water fishing
T	X			
BP		X		
R			X	
S				X
T/BP	X	X		
T/R	X		X	
T/S	X			X
BP/R		X	X	
BP/S		X		X
R/S			X	X
T/BP/R	X	X	X	
T/BP/S	X	X		X
T/R/S	X		X	X
BP/R/S		X	X	X
T/BP/R/S	X	X	X	X

estimation purposes in logistic form. Each equation in the simultaneous system simply says that the logarithm of the odds of doing "some" fishing of any one of four kinds is potentially influenced by the log of the odds of doing "some" of each of the other alternative activities in the set, by a group of socioeconomic variables, and by the physical water availability measure specific to the activity represented in the equation. This view of the full simultaneous conditional choice system is suggested by Mallar (1977). In logit equation form it may be written as in table 4-3. Freshwater availability is included in these equations in terms of the relative amounts of water suitable for trout, bass/pan and rough fishing by state. (The variables TW, BW, and RW are defined in table 4-3.) For the estimation exercise, we concentrate on acreage allocations by species class determined by our survey of state officials as described in chapter 2. The system is completed by inclusion of a measure of saltwater availability: a dummy variable that differentiates between states that have saltwater (or Great Lakes) coastlines and those that do not. We also use two alternative measures for these water availability variables as explained at some length in chapter 2 and summarized in table 3-3.

Our interest is in the reduced-form equations of the simultaneous model, however, rather than the structural equations themselves, because our objective is to predict the probabilities of doing some of the four (really three, since we are not dealing with changes in saltwater quality) types of fishing conditioned on post-policy values of the water availability variables. Thus, we are not concerned about the order conditions for econometric identification of the structural models.

The reduced form resulting from the four-equation simultaneous model can be shown to be comprised of four equations expressing each dependent variable (I_i) as a function of all the predetermined variables in the system; Z, TW, BW, RW, and C. Since TW = 1-BW-RW, we can write the reduced-form system in terms of the π_{ij} reduced-form parameters and the exogenous variables (now ignoring the saltwater equations) as:

Table 4-3. The Simultaneous Choice Model

<u>CHOICE</u>

Some cold-water fishing: $$I_1 = \log \frac{P(F_1=1 | F_2, F_3, F_4)}{P(F_1=0 | F_2, F_3, F_4)} = \beta_{11}Z + \beta_{12}TW + \gamma_{12}I_2 + \gamma_{13}I_3 + \gamma_{14}I_4$$

Some warm-water game fishing: $$I_2 = \log \frac{P(F_2=1 | F_1, F_3, F_4)}{P(F_2=0 | F_1, F_3, F_4)} = \beta_{21}Z + \beta_{22}BW + \gamma_{21}I_1 + \gamma_{23}I_3 + \gamma_{24}I_4$$

Some rough fishing: $$I_3 = \log \frac{P(F_3=1 | F_1, F_2, F_4)}{P(F_3=0 | F_1, F_3, F_4)} = \beta_{31}Z + \beta_{32}RW + \gamma_{31}I_1 + \gamma_{32}I_2 + \gamma_{32}I_4$$

Some saltwater fishing: $$I_4 = \log \frac{P(F_4=1 | F_1, F_2, F_3)}{P(F_4=0 | F_1, F_2, F_3)} = \beta_{41}Z + \beta_{42}C + \gamma_{41}I_1 + \gamma_{42}I_2 + \gamma_{43}I_3$$

where:

I_i = Index (log of the odds) of doing some fishing of type i

Z = Exogenous explanatory variables including income, age, sex, area of residence, etc. (Treated as a single variable above to simplify notation)

TW = Ratio of acres of water suitable for cold-water game fishing to total acres of fishable water in state of residence

BW = Ratio of acres of water suitable for warm-water game fishing to total acres of fishable water in state of residence

RW = Ratio of acres of water suitable for rough fish fishing to total acres of fishable water in state of residence

C = Indicator of accessibility of marine or Great Lakes fishing by state of residence

TW+BW+RW = 1.0

F_1 = Do some cold-water game fish fishing (1) or not (0)

F_2 = Do some warm-water game fish fishing (1) or not (0)

F_3 = Do some rough fish fishing (1) or not (0)

F_4 = Do some saltwater, estuarine or Great Lakes fishing (1) or not (0)

The Reduced-Form Choice Model

CHOICE	EQUATION
Some cold-water game fishing:	$I_1 = \pi_{11}Z + \pi_{12}BW + \pi_{13}RW + \pi_{14}C$
Some warm-water game fishing:	$I_2 = \pi_{21}Z + \pi_{22}BW + \pi_{23}RW + \pi_{24}C$
Some rough fishing:	$I_3 = \pi_{31}Z + \pi_{32}BW + \pi_{33}RW + \pi_{34}C$

The reduced-form model shows that if the simultaneous choice model is appropriate, the same set of exogenous variables must appear in every equation of the reduced-form choice model to be estimated.

If we posit a sequential rather than simultaneous decision process, we find that the conditional choice equations to be estimated contain the same set of arguments. Figure 4-1 shows such a sequential decision process. First, the individual decides whether to fish or not contingent on total water availability and then whether to do freshwater fishing or saltwater fishing contingent on saltwater availability. If he decides on some freshwater, he first decides whether to pursue only rough fish, only game fish, or some of both. If he decides on only game fish or some of both, he then must allocate his choices among cold-water game fish, warm-water game fish and rough fish. Every choice in the chain which is the subject of this chapter is influenced by the availability of saltwater and the relative availability of water suitable for cold-water game fishing, warm-water game fishing and rough fishing.

Independent and Dependent Variables in the Choice Equations

Individual choices about participation in specific types of fishing are influenced by other exogenous variables in addition to water availability, and we lumped these into a catchall, Z, above. The elements of Z in our models are socioeconomic variables and measures of expected catch (bag) in the several alternative activities. These variables were defined and the sources used for them summarized in chapter 3 (table 3-3), but it will be convenient to list them again here and to add a bit of information about the bag variable.

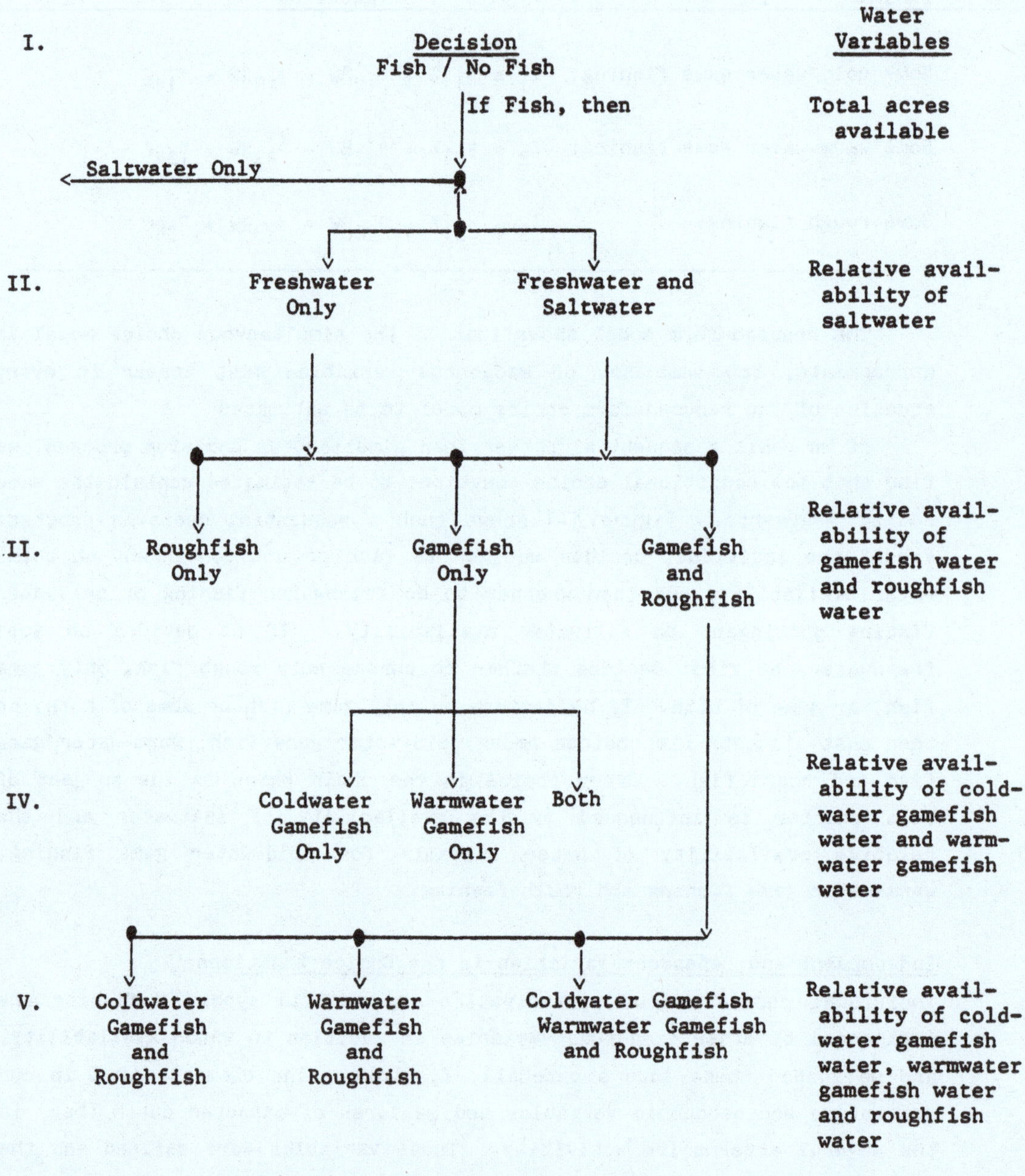

Figure 4-1. The sequential decision process

The socioeconomic variables included in our choice model are:

Household income in 1975 dollars

Respondent's age in years

Age squared

Respondent's sex

Respondent's residence - metropolitan or nonmetropolitan

All these variables were taken from the records of the National Survey of Fishing, Hunting, and Wildlife-Associated Recreation (NSHFWR) mail survey. (U.S. Department of the Interior, n.d.) (The sizes and compositions of our alternative samples of the universe of fisherman records are discussed below.)

Our measures for the quality of fishing--bags, by species, per day--are compromise proxies. The mail survey asked each respondent to report his favorite specific hunting or fishing activity (elk hunting, bass fishing, trout fishing, and the like) and the days spent doing it. Individuals were also asked their seasonal bag in this favorite activity, from which their average daily bag is easily calculated. The fishermen responding to this question are probably relatively expert, so that their reported catches are probably in the upper tail of the distribution of catches for all individuals in each category, and thus represent the best any fisherman could do. For some states very few observations were available in a category, even when the categories were defined broadly as cold-water game fish, warm-water game fish/panfish and rough fish. Also, some individuals reported what appeared to be incredible catches. To get reasonable measures, therefore, we aggregated average individual catches per season into the eleven regional categories reported in table 4-4. These become our explanatory "bag" variables in the equations reported below.[5]

The dependent variables fall into two groups as already indicated: a mutually exclusive set of choice categories and a nonmutually exclusive set which can be obtained by aggregation from the former. These variables were defined in table 3-3, but a reminder of their structures and names will be useful. The mutually exclusive category variables are dummy variables with a value of 1 if the respondent falls into that category and zero otherwise. The variable names are self-explanatory combinations of the letters T (trout fishing), BP (bass and pan fishing), R (rough fishing), and S (saltwater fishing) as shown in table 4-2. The nonmutually exclusive category variables are also dummies, taking the value of 1 if the

Table 4-4. Average Daily Catch for Favorite Activity by Activity and Region

Region[a]	Cold-water Game fish	Warm-Water Game fish, Panfish	Rough fish
1. Northwest	3.0	11.4	0.6
2. Pacific	2.7	3.3	3.2
3. South Mountain	2.6	4.6	1.7
4. North Mountain	2.5	7.7	1.3
5. West North Central	2.3	4.5	2.5
6. West South Central	2.4	13.2	0.9
7. East North Central	2.2	6.2	1.2
8. East South Central	1.4	10.1	4.2
9. South Atlantic	2.6	9.7	1.3
10. Middle Atlantic	2.8	3.8	0.2
11. New England	1.9	5.4	1.8

[a]Region-state key:

Region	States
1.	Washington, Oregon
2.	California
3.	Nevada, Utah, Colorado, Arizona, New Mexico
4.	Idaho, Montana, Wyoming
5.	North Dakota, South Dakota, Nebraska, Kansas, Minnesota, Iowa, Missouri
6.	Oklahoma, Arkansas, Louisiana, Texas
7.	Wisconsin, Illinois, Indiana, Michigan, Ohio
8.	Kentucky, Tennessee, Mississippi, Alabama
9.	Florida, Georgia, South Carolina, North Carolina, Virginia, West Virginia, Maryland, Delaware, Washington, D.C.
10.	Pennsylvania, New York, New Jersey
11.	Maine, Vermont, New Hampshire, Massachusetts, Connecticut, Rhode Island

respondent fell in <u>any</u> of the categories including that kind of fishing. The names combine the letter designations for type as above with MAN. Thus TMAN = 1 if the respondent did <u>any</u> trout fishing and zero otherwise.

Before turning to the estimation results, we discuss the sample design problems presented by the NSHFWR mail survey data, as these have a direct bearing on our preference for OLS over logit estimation of the conditional choice probabilities of doing some fishing in each of the nonmutually exclusive categories.

The Usable NSHFWR Mail Survey Data Set: Sample Design Problems and their Implications for Estimation Methods

The peculiarities of the telephone survey phase of NSHFWR discussed in chapter 3 carry over into the mail survey phase. The latter represents a subset of observations drawn from the telephone survey with a similar method of selecting an equal number of participants per state, and leading to the same underrepresentation of relatively populous states and overrepresentation of nonmetropolitan residents within each state.

To find the approximately correct rate of state representation of fishermen in the total population of fishermen and also the approximately correct urban-rural division of those fishermen, we created a very large (50,000 observation) proportional sample from the telephone survey based on census parameters using a procedure identical to that outlined in the previous chapter. Then, we identified people who fished (14,105) in this "proportionalized" telephone survey subsample and calculated the "representative" frequencies shown in table 4-5. Table 4-5 suggests that had the mail survey of recreationists been designed randomly, 67.5 percent of the responding fishermen would have lived in metropolitan areas and 32.5 percent would have lived in nonmetropolitan areas, distributed across the lower 48 states and the District of Columbia as shown in the second column of the table.

The mail survey component of NSHFWR contains information on over 20,000 individuals who enjoyed at least one of the two broad recreational activities covered in the questionnaire (hunting and fishing). Of these, 14,826 were fishermen. Metropolitan area residents accounted for 36.5 percent of the total fishermen (5,405 records), and nonmetropolitan area residents for 63.5 percent (9,241 records), almost exactly the reverse of the desired proportions. The way this distortion can be eliminated is discussed below. Another problem, but one we can do nothing about, is the

Table 4-5. Fisherman Frequencies in Proportional Sample of 50,000 Respondents from Telephone Survey

State	Number of fishermen	Relative frequency of all fishermen	Relative frequency of metropolitan fishermen	Relative frequency of nonmetropolitan fishermen
ALABAMA	316	0.022	0.01219	0.00981
ARIZONA	117	0.008	0.00520	0.00280
ARKANSAS	204	0.014	0.00480	0.00920
CALIFORNIA	1288	0.091	0.08308	0.00792
COLORADO	196	0.014	0.01128	0.00272
CONNECTICUT	154	0.011	0.00943	0.00157
DELAWARE	34	0.002	0.00171	0.00029
DISTRICT OF COLUMBIA	16	0.001	0.00113	0
FLORIDA	606	0.043	0.03384	0.00916
GEORGIA	364	0.026	0.01173	0.01427
IDAHO	82	0.006	0.00139	0.00461
ILLINOIS	634	0.045	0.03249	0.01251
INDIANA	423	0.030	0.01992	0.01008
IOWA	239	0.017	0.00605	0.01095
KANSAS	198	0.014	0.00546	0.00858
KENTUCKY	220	0.016	0.00560	0.01040
LOUISIANA	327	0.023	0.01456	0.00844
MAINE	76	0.005	0.00072	0.00428
MARYLAND	229	0.016	0.01293	0.00307
MASSACHUSETTS	290	0.021	0.01701	0.00399
MICHIGAN	681	0.048	0.03581	0.01219
MINNESOTA	376	0.027	0.01744	0.00956
MISSISSIPPI	191	0.014	0.00403	0.00997
MISSOURI	381	0.027	0.01596	0.01104
MONTANA	65	0.005	0.00123	0.00377
NEBRASKA	112	0.008	0.00378	0.00422
NEVADA	39	0.003	0.00246	0.00054
NEW HAMPSHIRE	47	0.003	0.00128	0.00172
NEW JERSEY	291	0.021	0.01848	0.00252
NEW MEXICO	58	0.004	0.00138	0.00262
NEW YORK	857	0.061	0.05008	0.01092
NORTH CAROLINA	354	0.025	0.01102	0.01398
NORTH DAKOTA	49	0.003	0.00055	0.00245
OHIO	699	0.050	0.03965	0.01035
OKLAHOMA	269	0.019	0.00988	0.00912
OREGON	184	0.013	0.00784	0.00516
PENNSYLVANIA	501	0.036	0.02736	0.00864
RHODE ISLAND	37	0.003	0.00276	0.00024
SOUTH CAROLINA	178	0.013	0.00621	0.00679
SOUTH DAKOTA	53	0.004	0.00060	0.00340
TENNESSEE	297	0.021	0.01336	0.00764
TEXAS	1023	0.073	0.05716	0.00584
UTAH	121	0.009	0.00723	0.00178
VERMONT	27	0.002	0	0.00191
VIRGINIA	322	0.023	0.01543	0.00757
WASHINGTON	358	0.025	0.01662	0.00838
WEST VIRGINIA	98	0.007	0.00293	0.00407
WISCONSIN	391	0.028	0.01562	0.01238
WYOMING	33	0.002	0	0.00234
TOTAL	14105	1.000	0.675	0.325

fact that the mail phase of NSHFWR had a response rate of only 37 percent and no correction for nonresponse bias was attempted. Therefore, there is always the possibility that the behavior of some segment of the population - perhaps individuals with limited reading and writing skills - may be omitted entirely in the equations we estimate.

The dependent and independent variable measures required to perform the estimation exercises reported in this chapter had to be constructed from several different answers to questions in the survey. When each of the 14,826 fishermen observations was scanned to create the variables for estimation, some with incomplete information had to be discarded, leaving a total of 11,168 observations with no missing data. The usable sample thus produced proved to be just as distorted as the initial 14,826-observation sample of fishermen.

Comparison of tables 4-5 and 4-6 shows how badly distorted this usable sample of observations from the mail survey is. Where a random sample would have given, on average, 67.5 percent of the total fisherman population from metropolitan areas, the usable mail survey records give only 36.8 percent. Underrepresentation of fishermen from populous states is another shortcoming of this survey sample. While we would expect California, New York, Ohio, and Texas, respectively, to have 9.1, 6.1, 5.0, and 7.3 percent of the total population in a proportional sample, each of these states commands only about 1.7 percent of the total fishermen in the usable mail survey sample.

These characteristics present serious problems for econometric estimation, since, based on our previous discussion in chapter 3, we have good reason to believe that the coefficients of a microdata participation model estimated from the nonproportional selective sample using the logit technique will be biased. Therefore, in logit estimation we cannot use all 11,168 mail survey observations from the mail sample. We can construct a perfectly proportional subsample by selecting the total sample size that would satisfy the most binding of the following constraints:

Table 4-6. Fisherman Frequencies for Usable Records in Mail Survey Sample

State	Total usable fisherman records	Usable metropolitan fisherman records	Usable nonmetro fisherman records	Relative frequency of all fishermen	Relative frequency of metro fishermen	Relative frequency of nonmetro fishermen
ALABAMA	178	62	116	0.016	0.00555	0.01039
ARIZONA	218	122	96	0.020	0.01092	0.00860
ARKANSAS	199	27	172	0.018	0.00242	0.01540
CALIFORNIA	199	155	44	0.018	0.01388	0.00394
COLORADO	288	139	149	0.026	0.01245	0.01334
CONNECTICUT	212	89	123	0.019	0.00797	0.01101
DELAWARE	219	105	114	0.020	0.00940	0.01021
DIST. OF COLUMBIA	4	4	0	0	0.00036	0
FLORIDA	203	123	80	0.018	0.01101	0.00716
GEORGIA	182	65	117	0.016	0.00582	0.01048
IDAHO	291	55	236	0.026	0.00492	0.02113
ILLINOIS	453	199	254	0.041	0.01782	0.02274
INDIANA	222	80	142	0.020	0.00716	0.01271
IOWA	211	72	139	0.019	0.00645	0.01245
KANSAS	220	56	164	0.020	0.00501	0.01468
KENTUCKY	186	36	150	0.017	0.00322	0.01343
LOUISIANA	176	95	81	0.016	0.00851	0.00725
MAINE	237	33	204	0.021	0.00295	0.01827
MARYLAND	165	102	63	0.015	0.00913	0.00564
MASSACHUSETTS	200	84	116	0.018	0.00752	0.01039
MICHIGAN	242	124	118	0.022	0.01110	0.01057
MINNESOTA	244	96	148	0.022	0.00860	0.01325
MISSISSIPPI	148	31	117	0.013	0.00278	0.01048
MISSOURI	193	50	143	0.017	0.00448	0.01280
MONTANA	303	82	221	0.027	0.00734	0.01979
NEBRASKA	227	62	165	0.020	0.00555	0.01477
NEVADA	247	104	143	0.022	0.00931	0.01280
NEW HAMPSHIRE	238	8	230	0.021	0.00072	0.02059
NEW JERSEY	209	134	75	0.019	0.01200	0.00672
NEW MEXICO	230	49	181	0.021	0.00439	0.01621
NEW YORK	197	112	85	0.018	0.01003	0.00761
NORTH CAROLINA	178	45	133	0.016	0.00403	0.01191
NORTH DAKOTA	230	33	197	0.021	0.00295	0.01764
OHIO	185	106	79	0.017	0.00949	0.00707
OKLAHOMA	194	106	88	0.017	0.00949	0.00788
OREGON	300	150	150	0.027	0.01343	0.01343
PENNSYLVANIA	234	111	123	0.021	0.00994	0.01101
RHODE ISLAND	192	114	78	0.017	0.01021	0.00698
SOUTH CAROLINA	190	84	106	0.017	0.00752	0.00949
SOUTH DAKOTA	219	18	201	0.020	0.00161	0.01800
TENNESSEE	162	39	123	0.015	0.00349	0.01101
TEXAS	188	67	121	0.017	0.00600	0.01083
UTAH	283	183	100	0.025	0.01639	0.00895
VERMONT	225	0	225	0.020	0	0.02015
VIRGINIA	193	81	112	0.017	0.00725	0.01003
WASHINGTON	666	251	415	0.060	0.02247	0.03716
WEST VIRGINIA	205	44	161	0.018	0.00394	0.01442
WISCONSIN	328	127	201	0.029	0.01137	0.01800
WYOMING	255	0	255	0.023	0	0.02283
Total	11168	4114	7054	1.000	0.36837	0.63163

$$N^* \equiv \min_{ij} \left[\frac{Q_{ij}}{\alpha_{ij}}\right]$$

Q_{ij} = available observations in mail sample

α_{ij} = proportional frequency implied for i^{th} state, j^{th} residence status from table 4-3. $\sum_i \sum_j \alpha_{ij} = 1.00$.

That is, N* is the total subsample size consistent with keeping all of its cells exactly in line with the desired relative frequencies given in table 4-5.

Unfortunately, the largest subsample that could be drawn from the usable mail survey data without violating the perfect proportionality constraints implied in table 4-5 is 1,172 observations. The variance of each of the parameter estimates in each of the logit equations is likely to be very large when working with such a small sample, so the resulting coefficient estimates are not likely to be very efficient.

There is no clean solution to this problem. If we want to perform logit estimation we must either accept the loss of information implied by reducing the sample size by 90 percent to achieve perfect proportionality or seek some compromise between perfect proportionality and sample size. One such compromise sample design can be produced by asking if there is a "best" subsample size greater than 1,172 but less than the available universe, given that the subsample increment of observations above 1,172 is allocated across state-residency status cells in an "optimal" way. One reasonable rule for "optimal" allocation is the minimization of the sum of squared deviations between the actual cell frequencies obtained and the targeted cell frequencies desired. This is a quadratic programming problem which involves the minimization of a quadratic criterion function subject to linear constraints. The mechanics of solving the problem are discussed in appendix B. Here it is sufficient to say that the solution yielded a "best quasi-proportional" sample of almost 5,000 observations which could be used with some confidence for logit estimation of the reduced forms set out above.

Another appealing alternative is to abandon logit estimation and opt instead for weighted least squares (WLS) estimation of a linear probability model of the reduced forms with 0, 1 binary dependent variables. The proper case weights, D_{ij}, which appear in the Ω^{-1} matrix of a GLS-like estimation procedure are (see appendix C):

$$\frac{\alpha_{ij}}{Q_{ij} / \sum_{i} \sum_{j} Q_{ij}}$$

The appeal of case-weighted least squares estimation of the conditional choice equations is that it allows us to take advantage of all the information in the 11,168 observation sample while simultaneously correcting for the bias its unproportional design imparts to the estimated regression coefficients. The disadvantages are, at least theoretically, legion: particularly that the error term of the linear probability model is heteroskedastic and that such models can predict outside the 0-1 probability interval (Pindyck and Rubinfeld, 1976, pp. 238-243). Those objections are, in our opinion, more matters of theoretical nicety than of practical difficulty. First, the fish/no fish OLS and logit results presented in the preceding chapter demonstrate no great divergence in terms of the relevant partial derivatives of the probability functions or in terms of statistical tests of significance. Furthermore, the problem of prediction outside the 0-1 interval is only serious for large changes in the independent variables away from their means, which is hardly the case in our data, as chapter 2 demonstrated.

In our opinion, these potential problems with WLS do not outweigh the advantages it offers in overcoming the definitely real problems of sample design and sample size in our application.[6] We do show how the key partial derivatives vary with sample size and estimation technique and in chapter 6 how the use of the logit technique and the quasi-proportional sample change the benefit estimates from those obtained via our principal route--WLS applied to the full available sample. We also show there the extent to which the combination of ordinary least squares and the quasi-proportional sample changes our findings. By way of preliminary evidence, table 4-7 shows how the samples themselves differ in terms of sample means of the dependent and independent variables. The major lesson of this table is that failure to weight the full (nonproportional) sample leaves a very different basis for estimation than is represented by any of the other

Table 4-7. Pairwise Comparisons of the Significance of the Difference in Variable Means Across Samples

Variable name	Variable definition	Means of variables: Small sample: N=1184	Medium sample: N=4990	Full sample: N=11168	Weighted full sample: N=11168	Absolute "t" values of significance of difference between means[a]: Small versus medium sample	Small versus full sample	Small versus full weighted sample	Medium versus full sample
DEPENDENT VARIABLES									
T	Mutually exclusive categories of fishing as combinations of cold-water game fish (T), warm-water game fish (B), panfish (P), cat-rough fish (R) or saltwater-Great Lakes (S) fishing. Sum over all categories = 1.00.	0.099	0.118	0.196	0.101	1.92*	10.31*	0.22	13.51*
B		0.032	0.030	0.026	0.030	0.32	1.20	0.37	1.60
P[b]		0.083	0.076	0.060	0.075	0.73	2.69*	0.95	3.65*
R		0.026	0.031	0.029	0.028	0.85	0.67	0.41	0.41
S		0.120	0.114	0.093	0.121	0.58	2.76*	0.10	3.99*
TB		0.017	0.014	0.022	0.014	0.60	1.39	0.77	3.67*
TP		0.023	0.021	0.026	0.020	0.45	0.79	0.66	2.29*
TR		0.012	0.010	0.016	0.008	0.41	1.33	1.23	3.14*
TS		0.037	0.045	0.052	0.046	1.24	2.54*	1.54	2.00*
BP		0.083	0.088	0.074	0.087	0.63	1.07	0.47	3.09*
BR		0.020	0.019	0.016	0.019	0.27	0.95	0.23	1.24
BS		0.018	0.014	0.009	0.020	0.84	2.09*	0.49	2.48*
PR		0.044	0.064	0.055	0.062	2.88*	1.82	2.82*	2.03*
PS		0.026	0.020	0.015	0.020	1.30	2.33*	1.24	2.01*
RS		0.007	0.007	0.005	0.008	0.05	0.63	0.40	1.06
TBP		0.042	0.030	0.037	0.031	1.89	0.86	1.81	2.23*
TBR		0.004	0.007	0.009	0.007	1.17	2.43*	1.48	1.69
TBS		0.007	0.008	0.009	0.008	0.61	0.76	0.40	0.17
TPR		0.023	0.023	0.023	0.016	0.03	0.07	1.56	0.18
TPS		0.014	0.011	0.010	0.012	0.68	0.97	0.28	0.51
TRS		0.007	0.005	0.005	0.007	0.52	0.70	0	0.32
BPR		0.071	0.083	0.074	0.080	1.38	0.34	1.14	1.94
BPS		0.037	0.029	0.020	0.033	1.32	3.04*	0.70	3.40*
BRS		0.005	0.005	0.004	0.008	0.11	0.69	0.90	1.09
PRS		0.019	0.019	0.013	0.020	0.20	1.38	0.48	2.89*
TBPR		0.026	0.025	0.031	0.023	0.14	0.92	0.62	1.90
TBPS		0.028	0.024	0.018	0.026	0.73	1.98*	0.40	2.38*
TBRS		0.004	0.004	0.004	0.006	0.09	0.10	0.99	0.34
TPRS		0.013	0.011	0.011	0.011	0.35	0.62	0.59	0.48
BPRS		0.032	0.023	0.019	0.029	1.59	2.57*	0.56	1.89
TBPRS		0.023	0.020	0.018	0.024	0.58	1.12	0.22	0.99
TMAN	People who do "some" cold-water game fish (TMAN), warm-water game fish (BMAN), pan fish (PMAN), cat-rough fish (CMAN) or salt-water (OMAN) fishing.	0.378	0.378	0.488	0.360	0.03	7.42*	1.22	13.19*
BMAN		0.449	0.426	0.389	0.446	1.47	3.98*	0.32	4.39*
PMAN		0.585	0.568	0.503	0.569	1.08	5.43*	1.06	7.65*
CMAN		0.335	0.356	0.332	0.357	1.40	0.25	1.45	3.07*
OMAN		0.395	0.361	0.304	0.398	2.19*	6.15*	0.20	7.03*
EXPLANATORY VARIABLES									
INCH	Household Income	15594	15300	14711	15797	0.78	2.50*	0.57	3.03*
AGE	Age	35.84	35.54	35.65	35.62	0.59	0.40	0.47	0.39
SEX	Fraction Female	0.142	0.142	0.147	0.141	0.02	0.43	0.09	0.84
METRO	Fraction Metropolitan	0.655	0.567	0.368	0.675	5.68*	19.73*	1.38	23.77*
COAST	Fraction with Sea/ Great Lakes Coast	0.80	0.725	0.59	0.796	5.35*	16.46*	0	17.35*
COLDACRE	Coldwater Acreage per Capita	0.033	0.056	0.131	0.037	5.52*	23.32*	0.84	22.11*
Number of Significantly Different Means						6	18	1	24

*Significant at the 5% level, two-tail test

[a] $t = (\bar{X}_1 - \bar{X}_2) / \left(\frac{S_1^2}{N_1} + \frac{S_2^2}{N_2} \right)^{1/2}$

where subscripts 1 and 2 refer to the two samples, and S^2 is the sample variance, and the population variances of the two samples are not assumed to be equal.

[b] Note that in this table panfishing (P) is separated from bass fishing in both the mutually exclusive and the inclusive dependent variables. This difference in treatment does not affect our conclusions about the a priori importance of sample differences.

three available samples: the smallest, perfectly proportional one; the medium, quasi-proportional one; or the full sample after weighting.

Results: Weighted OLS Equations

Before turning to the final set of equations, we first establish the link between the equations estimated on the first dependent variable set -- the mutually exclusive choice categories -- and the second dependent variable set -- the nonmutually exclusive choice categories. To do so we perform weighted single-equation least-squares estimation for the fifteen mutually exclusive choice categories introduced in table 4-2. The overarching constraint on the system of all fifteen equations is that the sum of the predicted cell probabilities must be one. In the estimation context, the restriction says that the sum of the intercept terms across all fifteen equations must equal one and that the across-equation sums of each of the slope parameters must equal zero (Pindyck and Rubinfeld, 1976, p. 280). The fifteen equations can be estimated singly rather than as a restricted system because, with the same set of explanatory variables appearing in each equation, the restriction is automatically satisfied (Kmenta, 1971, pp. 521-523).[7]

The parameters of the fifteen equations estimated using our state survey water-share-by-species-class variables appear in table 4-8. Inspection of the last row of the table (labelled "Column Sum of Coefficients") shows that the restriction implied by the mutually exclusive partitioning of the choice categories is satisfied. Also encouraging is the fact that the variable measuring the availability of saltwater/Great Lakes fishing (COAST) is significant for all equations, while the variables measuring the relative availability of warm-water/panfish water (RFFBPSHR) and rough fish water (RFFRSHR) are significant at the 5 percent level in eleven of fifteen and eight of fifteen equations, respectively.[8]

The coefficients in these equations confirm what is generally believed to be true in the sociology of angling about the distinction between "purists" and other anglers. In the four "purist" categories (T, BP, R, S) higher income is associated with an increase in the probability of doing exclusively cold-water game fish or saltwater fishing, skill-intensive pursuits which are traditionally believed to attract higher income individuals who can afford expensive outfits. Income appears not to be a factor influencing the choice of exclusive warm-water game fish/panfish angling and has a negative influence on the probability of an exclusive

Table 4-8. Weighted Least Squares (WLS) Estimation of Fifteen Mutually Exclusive Conditional Probabilities of Fishing Participation by Species Combination Sought, Given One is an Angler

(absolute value of "t" statistic in parentheses)

DEPENDENT VARIABLE		INTERCEPT	INCH	AGE	AGESQ	SEX	METPO	COAST	TBAG	BPBAG	CBAG	RFFBPSHR	RFFRSHR	R^2	F RATIO (11;9972 D.F.)
		INDEPENDENT VARIABLES													
1. T (Purist: only cold-water game fish)		0.4881 (20.27)*	0.0000009 (3.44)*	-0.0029 (2.86)*	0.000046 (3.77)*	0.0136 (1.61)	-0.0170 (2.58)*	-0.1157 (13.40)*	-0.0043 (1.24)	-0.0057 (7.90)*	0.0037 (1.61)	-0.2583 (20.32)*	-0.3341 (11.22)*	0.0864	85.74*
2. BP (Purist: only warm-water game fish/ panfish)		-0.0420 (1.31)	-0.0000003 (0.88)	0.0034 (2.52)*	-0.000010 (0.58)	-0.0323 (2.86)*	-0.0144 (1.64)	0.0327 (2.84)*	-0.0233 (5.02)*	-0.0008 (0.80)	0.0039 (1.28)	0.2056 (12.13)*	0.3058 (7.71)*	0.0375	35.29*
3. R (Purist: only warm-water rough fish)		0.0140 (1.09)	-0.0000003 (2.00)*	0.0004 (0.78)	-0.000003 (0.41)	0.0111 (2.45)*	-0.0119 (3.39)*	-0.036 (7.28)*	0.0004 (0.23)	0.0013 (3.44)*	-0.0004 (0.34)	0.0436 (6.42)*	0.0145 (0.91)	0.0164	15.10*
4. S (Purist: only salt-water or Great Lakes)		-0.0146 (0.57)	0.0000011 (4.17)*	0.0009 (0.82)	0.000012 (0.91)	0.0445 (5.03)*	0.0566 (8.23)*	0.1151 (12.75)*	0.0081 (2.22)*	0.0006 (0.78)	-0.0131 (5.44)*	-0.1190 (8.95)*	0.0098 (0.31)	0.0611	59.02*
5. TBF	Mixed Tastes	0.1634 (8.05)*	-0.0000002 (0.84)	0.0008 (0.90)	-0.00002 (1.96)*	-0.0044 (0.62)	-0.0143 (2.56)*	-0.0297 (4.07)*	0.0071 (2.41)*	-0.0047 (7.71)*	-0.0063 (3.26)*	-0.0358 (3.33)*	-0.0462 (1.84)	0.0154	14.14*
6. TR	Mixed Tastes	0.0411 (5.41)*	0.0000001 (0.85)	-0.0005 (1.51)	0.000005 (1.37)	0.0042 (1.59)	-0.0048 (2.31)*	-0.0098 (3.61)*	0.0004 (0.39)	-0.0004 (1.81)	-0.0001 (0.19)	-0.0110 (2.75)*	-0.0361 (3.84)*	0.0052	4.70*
7. TS	Mixed Tastes	0.0282 (1.64)	0.0000002 (1.10)	0.0008 (1.16)	-0.000009 (1.03)	-0.0003 (0.06)	0.0133 (3.90)*	0.0448 (7.24)*	0.0089 (3.55)*	-0.0006 (1.16)	0.0073 (4.44)*	-0.1049 (11.53)*	-0.0501 (2.35)*	0.0264	24.55*
8. BPR	Mixed Tastes	0.1307 (4.62)*	-0.000002 (5.15)*	-0.0021 (1.82)	0.000003 (0.23)	-0.0083 (0.84)	-0.0518 (6.70)*	-0.1362 (13.43)*	-0.0129 (3.16)*	0.0074 (8.71)*	0.0093 (3.44)*	0.2972 (19.90)*	0.2838 (8.12)*	0.1073	108.96*
9. BPS	Mixed Tastes	-0.0067 (0.32)	0.000001 (5.24)*	0.0010 (1.19)	-0.000011 (1.04)	-0.0441 (6.07)*	0.0230 (4.07)*	0.0676 (9.10)*	-0.0081 (2.71)*	0.0056 (9.04)*	-0.0056 (2.85)*	-0.0476 (4.35)*	-0.0443 (1.73)	0.0369	34.70*
10. RS	Mixed Tastes	-0.0018 (0.27)	-0.0000001 (1.48)	0.0003 (1.08)	-0.000005 (1.44)	0.0018 (0.75)	0.0007 (0.39)	0.0075 (3.07)*	0.0003 (0.33)	-0.0002 (1.04)	-0.0004 (0.57)	-0.0002 (0.04)	0.0148 (1.76)	0.0021	1.95*
11. TBPR	Mixed Tastes	0.1471 (8.75)*	-0.0000004 (2.68)*	-0.0013 (1.92)	0.000008 (0.93)	-0.0114 (1.92)	-0.0242 (5.27)*	-0.0523 (8.66)*	0.0067 (2.74)*	-0.0019 (3.75)*	-0.0080 (4.96)*	0.0132 (1.49)	0.0299 (1.44)	0.0237	22.02*
12. TBPS	Mixed Tastes	0.0432 (2.53)*	0.0000002 (1.21)	0.0007 (1.04)	-0.000017 (1.95)	-0.0202 (3.38)*	0.0068 (1.46)	0.0369 (6.04)*	0.0041 (1.67)	-0.0017 (3.31)*	-0.0038 (2.33)*	-0.0273 (3.04)*	-0.0598 (2.84)*	0.0151	13.86*
13. TRS	Mixed Tastes	0.0052 (0.72)	≈ 0 (0.46)	-0.0003 (1.10)	0.000002 (0.44)	0.0024 (0.94)	0.0058 (2.93)*	0.0071 (2.74)*	0.0017 (1.59)	-0.0006 (2.94)*	0.0032 (4.61)	-0.0016 (0.43)	-0.0185 (2.07)*	0.0084	7.65*
14. BPRS	Mixed Tastes	-0.0343 (1.85)	-0.0000006 (3.16)*	≈ 0 (0.08)	0.000008 (0.85)	0.0094 (1.45)	0.0151 (2.99)*	0.0394 (5.95)*	0.0019 (0.70)	0.0039 (6.96)*	0.0025 (1.44)	0.0468 (4.80)*	0.0084 (0.37)	0.0169	15.58*
15. TBPRS		0.0382 (2.33)*	-0.0000002 (0.96)	-0.0011 (1.60)	0.000006 (0.73)	-0.0305 (5.31)*	0.0119 (2.66)*	0.0285 (4.87)*	0.0091 (3.86)*	-0.0021 (4.40)*	0.0078 (4.96)*	-0.0006 (0.07)	-0.0779 (3.85)*	0.0196	18.17*
Column sum of Coefficients		1.00	0	0	0	0	0	0	0	0	0	0	0		

Note: An asterisk (*) indicates significance at the 5 percent level.

rough fish choice. Likewise, purists, as is commonly maintained, seem to be relatively unconcerned about bag. At least we obtain insignificant coefficients on the bag variables particular to each purist activity, although it seems that potential purists are less likely to specialize if bags in alternative pursuits are high. Another interesting feature of all four purist equations is the positive (and significant for all but trout category) coefficient on SEX, suggesting that women who fish tend to be attracted to specialized rather than diversified pursuits.

Our intention is not, however, to explore the motivations behind particular types of fishing choices in detail. Instead, we aim to produce usable prediction equations for the choices which reflect the influence of water availability. The equations presented in table 4-9 are part of the road to this goal, and have been estimated using the nonmutually exclusive categories. These equations are fully consistent with those for the mutually exclusive categories (table 4-8) in the sense that each parameter estimate on any particular independent variable in an aggregate "some of" equation represents the sum of the corresponding parameter estimates from the appropriate disaggregated mutually exclusive equations. For example, in table 4-9 the coefficient attached to the RFFBPSHR variable in the "Some Trout Fishing" equation represents the sum of the coefficients in table 4-8 for the same variable across the T, TBP, TR, TS, TBPR, TRS, and TBPRS equations. The two alternative sets of equations are thus consistent with each other in the OLS framework.

It is difficult and perhaps not worthwhile to interpret the signs and magnitudes of the parameter estimates attached to the socioeconomic variables in table 4-11, because they are amalgams of the coefficients attached to the same variables in the mutually exclusive category regressions which have already been discussed. However, the parameter estimates for bag and acreage variables are easier to review. Specifically, we would expect that the choice of doing some fishing in any category should be positively related to that category's expected catch and its share of total fishable freshwater acreage available. Further, the availability of saltwater alternatives should have a dampening effect (that is, the coefficient should have a negative sign) on the probability

Table 4-9. Equations for the Prediction of Participation Probabilities for "Some of" Fishing Categories (Estimated by Weighted Least Squares, Using Population Weighted 11,000 Sample)

Variable	Some trout fishing	Some bass/pan fishing	Some rough fishing
INTERCEPT	0.9547 (25.52)*	0.4399 (12.12)*	0.3401 (9.17)*
INCH	0.0000006 (1.45)	-0.0000020 (4.98)*	-0.0000032 (7.83)*
AGE	-0.0038 (2.42)*	0.0013 (0.83)	-0.0047 (3.05)*
AGESQ	0.000021 (1.11)	-0.000048 (2.61)*	0.000009 (0.46)
SEX	-0.0466 (3.54)*	-0.0773 (6.07)*	-0.0213 (1.64)
METRO	-0.0174 (1.70)	-0.0479 (4.82)*	-0.0592 (5.84)*
COAST	-0.0901 (6.72)*	-0.0129 (0.99)	-0.1517 (11.41)*
TBAG	0.0336 (6.21)*	-0.0155 (2.94)*	0.0076 (11.42)*
BPBAG	-0.0178 (15.86)*	0.0056 (5.19)*	0.0073 (6.53)*
CBAG	0.0037 (1.04)	-0.0002 (0.05)	0.0139 (3.92)*
RFFBPSHR[a]	-0.4264 (21.58)*	0.4516 (23.56)*	0.3874 (19.79)*
RFFRSHR[a]	0.5927 (12.81)*	0.3996 (8.90)*	0.2190 (4.78)*
Degrees of freedom	9974	9972	9972
F ratio	100.93*	91.66	100.83*
R^2	0.0835	0.0918	0.1001

Note: An asterisk (*) indicates significance at the 5 percent level.

[a] Shares based on RFF survey of state officials (see chapter 2).

of the freshwater choices. These expectations are all fulfilled by the estimated equations.[9]

Table 4-10 shows how changing the basis for the water share variables affects the coefficients of those share variables. (All keep the same sign and remain strongly significant. We do not bother to give the t-statistics.) There are some large differences, though none of the alternatives is more than about twice as large or less than half as large.

We do not pursue further the influence of alternative measures of water availabiltiy on benefit estimates in the base case. There is, however, one modestly interesting feature of the regression models themselves. It concerns the coefficients on water-type shares, and its discussion requires a brief look back at table 4-9. There, in the trout equation, the coefficients on the bass/pan and rough water shares are both negative. This is to be expected, and using the identity that the sum of the shares is equal to one, these coefficients can be translated into a corresponding pair to attach to the trout and bass/pan water shares in an intuitively more pleasing structure. For the trout equation in table 4-9, the coefficients would be:

on trout water share: 0.593

on bass/pan water share: 0.166

The trout water share coefficient is simply the negative of the rough water share coefficient from the original equation, while the new bass/pan share coefficient is the difference between the original bass/pan share and the original rough water share coefficients.

When this is attempted with the equations estimated from the two synthetic sources, however, we find that the original share coefficients are not significantly different from each other (as measured by the F-test for coefficient equality). Under these water share measures, then, the trout choice model could have been estimated using the hypothesis that trout fisherman look only at the cold-water share of fishable waters in choosing whether or not to do any trout fishing. This hypothesis is not the one suggested by the heuristic choice models set out early in this chapter. An analogous contrast is to be found in the equations for bass/pan and rough fishing choices. That is, using the water shares from the survey of state officials, the hypothesis of interest in all shares is

Table 4-10. Contrasting Regression Results for Key Water Availability Variables and Equation Overall, Using Alternative Water Share Measures

	Water quality network model with Survey of Hatchery Fish		Water quality network model alone	
	Bass/pan share	Rough share	Bass/pan share	Rough share
Trout equation				
Ratio $\hat{B}/\hat{B}$ survey	2.18	1.67	1.18	0.87
F-Ratio	297.7		198.0	
R^2	0.230		0.166	
Bass/pan equation				
Ratio $\hat{B}/\hat{B}$ survey	1.54	2.09	0.75	0.86
F-Ratio	173.3		96.4	
R^2	0.148		0.088	
Rough fish equation				
Ratio $\hat{B}/\hat{B}$ survey	0.93	1.71	0.50	0.91
F-Ratio	91.0		103.6	
R^2	0.084		0.094	

not contradicted, while using the synthetic sources of water shares, the only breakdown of interest is cold versus warm (rough plus game) fishing water.

Before turning to the angler's choice of the number of days to fish each year, there is one other matter to pursue in connection with the probability equations; that is the impact of alternative choices of sample size and estimation method on key partial derivative estimates. To this end, table 4-11 shows how the partial derivatives of the water share variables compare for five combinations of sample size and estimation technique. The differences--here expressed as ratios to a primary combination, the 11,000 observation sample, and weighted least squares--are not dramatic, though again the highest is about 2 and the lowest about one half. The largest range in the ratios--the widest range of differences from too small to too large--is seen in the 1,100 observation logit model, with the 1,100 OLS model not far behind. The smallest range of ratios is seen in the results for OLS estimation on the 5,000 observation, quasi-proportional sample.

We return to the implications of these differences in chapter 6 when we report on variations in benefit estimates.

DATA AND RESULTS ON DAYS OF FISHING PARTICIPATION SPENT PER SEASON DIFFERENTIATED BY SPECIES SOUGHT

Dependent and Independent Variables

Each fisherman who participates in one or more of the nonmutually exclusive activity categories defined in the last section by definition devotes "some" days per year to that activity. The next major question is: How many days? And, even more important: How does the number of days change with a change in availability of fishable water of the relevant types?

The theoretical basis for choosing a model structure here is the same one already laid out for the probability-of-participation equations because this is simply the third step in the chain of decisions that result in particular quantities of recreation days (of particular types), given the availability of recreational resources. The arguments in each of these days-of-participation equations include, then, the usual socioeconomic variables and some measure of the quantity of water available, at least the quantity particular to that category (e.g., trout water availability in the trout equation).

Table 4-11. Comparison of Key Policy Partial Derivatives for Different Samples and Estimation Techniques

		Probability equations and variables					
		Some trout fishing		Some bass/pan fishing		Some rough fishing	
		Bass/pan share	Rough share	Bass/pan share	Rough share	Bass/pan share	Rough share
Sample Size	Estimating Method						
11000	WLS						
I	Partials	-0.43	-0.59	0.45	0.40	0.39	0.22
1100	OLS						
	Partials	-0.47	-0.35	0.32	0.48	0.37	0.40
	Ratios to I	1.09	0.59	0.71	1.20	0.95	1.82
1100	Logit						
	Partials	-0.51	-0.32	0.30	0.50	0.48	0.46
	Ratios to I	1.19	0.54	0.67	1.25	1.23	2.09
5000	OLS						
	Partials	-0.57	-0.50	0.49	0.50	0.38	0.30
	Ratios to I	1.32	0.85	1.09	1.25	0.97	1.36
5000	Logit						
	Partials	-0.60	-0.51	0.47	0.50	0.48	0.38
	Ratios to I	1.40	0.86	1.04	1.25	1.23	1.73

The equations below contain the same independent socioeconomic variables that were employed earlier in this chapter for estimating the second-step equations--with an important addition. We hypothesize that individuals who express strong preference for any kind of fishing activity over hunting or other wildlife-associated recreational activities would be more likely to spend a larger fraction of their leisure time fishing than otherwise. Therefore, the list of explanatory variables in Step III is broadened to include such a measure, called PREF, which equals 1 if individuals prefer fishing in general to other activities and 0 otherwise.

When it comes to specifying the relevant supply variables in each of the days-of-participation equations, another problem arises. It is whether we should include in each equation only the measures of catch and water availability peculiar to that category of participation, or whether we should bring into each equation all measures of catch and water availability. The latter alternative is hinted at, but not fully developed, by Cicchetti (1973, p. 37) who writes a days-of-participation equation for a particular activity as a function of socioeconomic variables, supply per capita, and the days spent in all other categories of recreational activities. This is a simultaneous model. If reexpressed as a set of reduced-form equations, measures of the supply of recreational facilities (or acreages in our case) in all activity categories would appear as arguments in the days equation particular to any activity category. The argument here is, then, analogous to that made for the specification of our Step II equations above. The counter argument is that the supply and catch variables not specific to a particular activity have no bearing on the days spent doing it. We test this alternative by an F test.

The actual measures of supply availability used are the acres of water per capita that are suitable for the relevant angling activities (e.g., trout water acres per capita) in the state. At this point we abandon the synthetic measures of pre-policy availability because we feel that the coverage problems of the water quality network model and other problems are too severe. That is, while the estimates of relative shares produced by the model, or the model augmented by Hatchery Fish (USDI, 1968), might not be too badly affected by coverage problems, the more demanding absolute measure, quantity per person, was not reliably measured.

The means of the dependent and independent variables for the three equations required in Step III appear in table 4-12. These means were

Table 4-12. Means of Days-of-Participation Equations Variables

Subsample	TROUTDA > 0	BAPADA > 0	ROUGHDA > 0
	Large source sample[a]		
VARIABLE			
I. Dependent			
TROUTDA	13.76	0	0
BAPADA	0	18.22	0
ROUGHDA	0	0	10.14
II. Independent			
AGE	34.45	34.53	32.48
AGESQ	1,408.63	1,410.64	1,260.18
PREF	0.34	0.37	0.34
SEX	0.13	0.13	0.13
METRO	0.69	0.66	0.61
INCH	15,606.68	15,446.21	14,011.89
COAST	0.79	0.81	0.74
TBAG	2.51	2.36	2.36
BPBAG	6.27	7.38	7.53
CBAG	1.58	1.60	1.72
RFFTRAC	0.077	0.030	0.023
RFFBPAC	0.081	0.119	0.111
RFFRAC	0.023	0.024	0.023
No. of Cases	4965	6066	3233

[a]Large unproportional sample. Means weighted by proportionalized state and metropolitan - nonmetropolitan residence status of fisherman (see appendix C).

computed from our large, unproportional mail survey sample on a population-weighted basis.

Results

The parameters of the days-of-participation equations estimated from the large unproportional sample using WLS appear in table 4-13. None of these equations fits the data particularly well in terms of the coefficient of determination, although we reject the hypothesis of no linear association.

Before discussing particular parameter estimates, we first want to ask whether the reduced-form equivalent of a fully simultaneous days-of-participation model, which implies that every acreage and bag variable appears in every reduced-form days equation, is more appropriate than the null hypothesis that only the acreage and bag variables peculiar to the dependent variable appear in the reduced-form to be estimated. The F test results for the comparison of full and restricted models are presented in table 4-14. Based on this large-sample regression, we would reject the null hypothesis for all three reduced models at the 5 percent level and for two of three at the 1 percent level. Thus, it appears to be inappropriate to exclude the bag and acreage variables not particular to the days equation being estimated. However, in evaluating the benefits of alternative water pollution control policies in chapter 6, we show how the restriction on exogenous variables affects our final answers.[10]

A closer look at the days equations (especially the unrestricted versions) reveals only one surprise. While the coefficients of the activity-specific acreage and catch variables in the cold-water game fish and warm-water game fish/panfish equations all have positive signs and are statistically significant, in the rough fish equation, the coefficients on rough fish acreage per capita and rough fish catch per day are significantly negative.

It is true that the days equations are reduced forms, and thus that their coefficients are nonlinear combinations of the coefficients in the underlying structural model, so that unless one is willing to say something a priori about the sign and magnitudes of these structural coefficients, one cannot really expect any particular sign on the reduced-form coefficients. But, intuition still strongly suggests that the coefficients on rough fish bag and acreage of rough fish water per capita should be positive in the rough fish days equation.

Table 4-13. Days of Participation Per Season Equations by Species Sought Estimated by WLS from the Large Unproportional Mail Survey Sample

("t" statistics in parentheses)

	Unrestricted			Restricted		
DEP VAR:	TROUTDA	BAPADA	ROUGHDA	TROUTDA	BAPADA	ROUGHDA
INTERCEPT	9.8432 (4.75)*	18.0586 (7.16)*	18.7881 (8.46)*	6.8941 (3.64)*	14.3915 (6.60)*	17.4440 (9.38)*
AGE	0.2685 (2.76)*	0.1480 (1.31)	-0.2651 (2.64)*	0.2957 (3.05)*	0.1520 (1.35)	-0.2582 (2.56)*
PREF	0.8460 (1.38)	3.8754 (5.62)*	0.2887 (0.46)	0.8721 (1.43)*	3.9046 (5.66)*	0.1757 (0.28)
AGESQ	-0.0018 (1.53)	-0.0017 (1.22)	0.0044 (3.43)*	-0.0021 (1.77)	-0.0017 (1.25)	0.0043 (3.37)*
SEX	-2.9101 (3.41)*	-7.7656 (7.74)*	-0.5118 (0.58)	-2.8299 (3.32)*	-7.7020 (7.68)*	-0.4588 (0.52)
METRO	-1.7851 (2.74)*	-1.3780 (1.92)	-0.5429 (0.86)	-1.2274 (1.92)	-1.4362 (2.01)*	-0.8791 (1.41)
INCH	-0.00013 (5.07)*	-0.00020 (6.51)*	-0.0001 (3.38)*	-0.00014 (5.40)*	-0.0020 (6.56)*	-0.00009 (3.17)*
COAST	0.6808 (0.92)	1.1291 (1.25)	-1.6070 (2.19)*	1.1421 (1.60)	1.4120 (1.65)	-1.4923 (2.04)*
TBAG	0.5257 (2.41)*	-1.2428 (2.95)*	-1.0774 (2.39)*	0.5874 (2.70)*	...	...
RFFTRAC	7.3370 (4.16)*	-1.9252 (0.56)	-6.9030 (1.91)	5.5138 (3.36)*	...	...
BBAG	-0.1034 (1.29)	0.3325 (4.39)*	0.2212 (2.92)*	...	0.3468 (4.60)*	...
RFFBPAC	-12.7414 (3.98)*	9.7593 (2.54)*	0.5182 (0.14)	...	7.4694 (3.09)*	...
CBAG	-0.0962 (0.37)	-0.1891 (0.76)	-0.6107 (3.55)*	...	...	-0.5355 (3.20)*
RFFRAC	1.2088 (0.21)	-11.1854 (0.82)	-28.1259 (2.19)*	...	...	-21.3090 (2.37)*
R^2 F	0.0246 9.62*	0.0275 13.18*	0.0240 6.09*	0.0205 11.54*	0.0259 17.87*	0.0175 (6.38)*

Note: An asterisk (*) indicates significance at the 5 percent level.

Table 4-14. F Tests of Unrestricted versus Restricted Days-of-Participation Models

Sample		R^2	n	p	q	Calculated F
Large	Unrestricted TROUTDA	0.0246	4965	13	...	5.20*
	Restricted TROUTDA	0.0205	4965	...	9	
	Unrestricted BAPADA	0.0275	6066	13	...	2.49*
	Restricted BAPADA	0.0259	6066	...	9	
	Unrestricted ROUGHDA	0.0240	3233	13	...	5.36*
	Restricted ROUGHDA	0.0175	3233	...	9	

<u>Note</u>: An asterisk indicates significance at the 5 percent level. F statistic calculated as:

$$F(p-q; n-p-1) = \frac{R_p^2 - R_q^2 / p-q}{(1-R_p^2) / n-p-1}$$

where

p is the number of parameters (excluding the intercept) in the full model; q the number of parameters in the reduced model; and n the number of observations.

These unexpected signs led us to experiment with the rough fish days equations along several lines, only one of which led to interesting results. We first speculated that southern behavior would differ from behavior in the rest of the country. However, inclusion of a southern dummy variable interacted with rough fish acreage and bag did not produce significant coefficients. We also speculated that everyone might prefer to pursue rough fish (especially catfish) in game fish quality water, and specified the relevant water availability variable in the equation to be only warm game fish water. However, the zero restrictions of this model could not be accepted on the basis of an F test against the specification of the full model, which includes all acreage and bag measures in every equation.

A final interesting possibility is that there are outlying observations in the data set which tend to mask the true relationship between days spent and the explanatory variables available in the survey. Simple descriptive statistics calculated from the large sample illustrate the potential outlier problem:

Days Per Season	TROUTDA	BAPADA	CATDA
Median	10	10	5
Maximum	365	325	188
Minimum	0.50	0.34	0.10
95th percentile	50	60	33

Note: Calculated from subsamples with nonzero days spent in each category.

The possible reasons for these outliers, especially those reporting days spent beyond, say the 95th percentile, are legion. All such observations cannot be regarded as "noise" or illegitimate data, since some retired or unemployed individuals with a passion for fishing might well spend more than a month or two per year fishing, especially if the climate is temperate and fishable water is close at hand. Unfortunately, we have no independent variables such as employment status which might explain such behavior. But, some of the data might be erroneous because of coding errors or the possibility, suggested by Hay and McConnell (1981), that some individuals could not recall the days they spent recreating with any accuracy, since the survey asked them about the intensity of their activity in the previous year.[11] Hay and McConnell (1981) corrected for the "noise" introduced by outliers on the high extreme by eliminating observations from their sample of nonconsumptive recreators (wildlife watching or photography) if the reporting individual claimed he spent more than twenty days per year. An idea of the salubrious effect of deleting outliers beyond the 95th percentile on the parameters of our days equations can be obtained by comparing the parameter estimates of table 4-15, produced from the large sample after deletion of outliers, with those in table 4-13.

Outlier deletion almost doubles R^2 in each equation and, more important, diminishes, though it does not eliminate, the unexpected negative effect of increased rough fish acreage on days spent rough fishing by nearly 70 percent. Similarly, days spent rough fishing still appear to be significantly affected in the negative direction by expected catch, although outlier deletion reduces the size of this effect by 77 percent.

The results of tables 4-13 and 4-15 suggest that rough fishing may be mainly a food-gathering exercise wherein individuals have a catch target which they attempt to meet by spending the minimum number of days in the field necessary. We can tell no more plausible story to explain the result obtained and feel that any attempt to do so would strain the reader's credulity. We do, however, show how the deletion of outliers from the days of participation sample affects our ultimate benefit figures by using the equations of both table 4-13 and 4-15 in the valuation exercise in chapter 6.

This completes our presentation of the participation stages of the benefit estimation method. In the next chapter we develop values for the days of participation in freshwater recreational fishing, differentiated by species class.

Table 4-15. Days of Participation per Season Equations by Species Sought Estimated by WLS from the Large Unproportional Mail Survey Sample after Outlier Deletion

DEP VAR:	Parameter estimates TROUTDA	BAPADA	ROUGHDA
INTERCEPT	8.3626 (7.75)*	12.3758 (9.40)*	9.7833 (11.24)*
AGE	0.2241 (4.46)*	0.2084 (3.54)*	-0.0198 (0.50)
PREF	0.7645 (2.40)*	1.9462 (5.36)*	0.2589 (1.04)
AGESQ	-0.0023 (3.73)*	-0.0019 (2.62)*	0.00066 (1.30)
SEX	-2.9643 (6.66)*	-4.5193 (8.71)*	-0.9890 (2.88)*
METRO	-0.9405 (2.76)*	-0.8394 (2.33)*	-0.9793 (3.96)*
INCH	-0.00008 (6.14)*	-0.00010 (6.31)*	-0.00006 (5.60)*
COAST	-0.0316 (0.08)	-0.3340 (0.71)	-1.2177 (4.25)*
TBAG	0.2576 (2.26)*	-0.7088 (3.28)*	-0.4990 (2.89)*
RFFTRAC	7.0472 (7.67)*	-4.4004 (2.45)*	-4.1686 (2.96)*
BBAG	-0.0765 (1.84)	0.0980 (2.38)*	0.0922 (3.14)*
RFFBPAC	-8.6298 (5.21)*	7.5352 (3.77)*	2.0436 (1.46)
CBAG	0.1712 (1.22)	-0.3289 (2.62)*	-0.1401 (2.11)*
RFFRAC	-1.2471 (0.42)	-7.3069 (1.03)	-8.8942 (1.80)
R^2	0.0417	0.0345	0.0471
F	15.87*	15.87*	11.63*
Cases deleted	213 of 4965	274 of 6066	161 of 3232
Percent deleted	4.3	4.5	5.0

Note: An asterisk (*) indicates significance at the 5 percent level.

NOTES

1. There are other divisions and choices as well. For example, the angler may choose among equipment types: jigging, bait casting, spin casting, spinning, and fly casting, are recognizably different ways of fishing, each of which has its capabilities and limitations in relation to type of water to be fished, type of fish sought, and even time of year. Because state fishery regulations often also specify that some pieces of water are reserved for particular methods, the links between water availability and type of equipment chosen are complex. But this and other divisions are of little interest in relation to our objective of benefit estimation, and we ignore them.

2. A similar contrast might be drawn for saltwater fishing but since our interest is entirely with freshwater, we shall let the opportunity pass.

3. Notice, however, that the proportion of Montanans with licenses is larger than the proportion claiming to be fishermen. We have no firm explanation for this unexpected result, but expect it is either an artifact of the survey sample or a real phenonmenon arising from misrepresentation of state of residence when purchasing a license.

4. We did not try to bring the Great Lakes fishing into our participation estimates. Because this is freshwater fishing and because at least the lowest lakes (Erie and Ontario) are likely to experience some change in quality after full implementation of FWPCA(A)/CWA, we are thus building in some degree of benefit underestimation. But none of the lakes is now unfishable and indeed only small parts of individual lakes could even to said to be confined to rough species. Therefore, the underestimation will be very small.

5. Except for individuals who responded to the bag question, for whom we used the reported bag for the appropriate bag category observation.

6. See the comments in Amemiya (1981). Also, it is worth quoting at length on this point from Deyak and Smith (1978, p. 73):

> A number of authors have questioned the applicability of OLS to models with dichotomous dependent variables, citing a variety of problems. A selected list of these includes: the potential for heteroskedasticity...; the failure of the estimated probabilities to be confined to the zero to one interval; and the problems with interpretation of the results of test statistics for null hypotheses associated with individual coefficients and sets of them. In each case, the evidence available in the literature suggests that these problems may not be great enough to favor an alternative method. Sampling experiments have suggested that the bias in the OLS variance-covariance matrix for the estimated

from the use of OLS versus nonlinear alternatives (i.e., logit and probit) are not sufficient to favor one technique over the others. This conclusion is further supported by the evidence on the technique's performance in hypothesis testing. Thus, since OLS seems quite robust and offers ready interpretations for the estimated coefficients, we have utilized it in our analysis.

It came to our attention after this study was completed that we could have performed logit analysis on the full sample using a weighted log-likelihood function, as explained in Manski and Lerman (1977).

7. This is simple to demonstrate in the case of a two-equation linear model with a single explanatory variable. Let $Y_1 = 1$ if $Y_2 = 0$ and $Y_2 = 1$, if $Y_1 = 0$ so $Y_2 = 1 - Y_1$. Then we have, for $\hat{Y}_i = \hat{\alpha}_i + \hat{\beta}_i X_i$ the familiar OLS formulas:

$$\hat{\beta}_1 = \frac{N\Sigma X_i Y_{1i} - \Sigma X_i \Sigma Y_{1i}}{N\Sigma X_i^2 - \left(\Sigma X_i\right)^2} \quad (1); \qquad \hat{\beta}_2 = \frac{N\Sigma X_i Y_{2i} - \Sigma X_i \Sigma Y_{2i}}{N\Sigma X_i^2 - \left(\Sigma X_i\right)^2} \tag{2}$$

Substituting $1 - Y_1$ in (2) and simplifying we find that:

$$-\hat{\beta}_2 = \hat{\beta}_1 \text{ or that } \hat{\beta}_2 + \hat{\beta}_1 = 0 \tag{3}$$

Similarly, for the intercept parameters:

$$\hat{\alpha}_1 = \frac{\Sigma Y_{1i}}{N} - \hat{\beta}_1 \frac{\Sigma X_i}{N} \tag{4}$$

$$\hat{\alpha}_2 = \frac{\Sigma Y_{2i}}{N} - \hat{\beta}_2 \frac{\Sigma X_i}{N} \tag{5}$$

Summing $\hat{\alpha}_1 + \hat{\alpha}_2$ we have:

$$\hat{\alpha}_1 + \hat{\alpha}_2 = \frac{\Sigma Y_{1i} + \Sigma Y_{2i}}{N} - \hat{\beta}_1 \frac{\Sigma X_i}{N} + \hat{\beta}_1 \frac{\Sigma X_i}{N} \tag{6}$$

or, since $\Sigma Y_{1i} + \Sigma Y_{2i} = N$ and the last two terms (6) cancel:

$$\hat{\alpha}_1 + \hat{\alpha}_2 = 1 \tag{7}$$

Taken together, (3) and (7) say that we can always solve for the parameters of the nth equation in the system from knowledge of the estimated parameters of the remaining n-1 equations, provided the same X_i appear in every equation.

8. Where j percentage variables sum to 100, only j-1 can be entered in a regression equation. In our equations the share of trout water has been left out.

9. Using other measures of water availability shares can produce some unexpected results. For example, when the water quality network model is the basis for share derivation, the coefficient on bass/pan bag is not significant in the WLS probability equation.

10. Further, when the same equations are estimated using the 5,000-observation, quasi-proportional sample and OLS, the patterns of coefficient signs and significance levels and, indeed, the relative and absolute sizes of the coefficients on all the key variables (water availability and catch) are very similar to those in table 4-13. It does not seem worthwhile to lengthen an already long story by reporting those regression results here. We do, however, include this variant in our examination of alternative benefit measures in chapter 6.

11. This is not a severe problem for the probability equations, since it is almost certainly easier for individuals accurately to recall what they did (fishing and what kind of fishing) than how often they did it.

REFERENCES

Amemiya, Takeshi. 1981. "Qualitative Response Models: A Survey," Journal of Economic Literature vol. 19 (December) pp. 1483-1536.

Cicchetti, Charles J. 1973. Forecasting Recreation in the United States (Lexington, Mass., Lexington Books).

Deyak, Timothy A. and V. Kerry Smith. 1978. "Congestion and Participation in Outdoor Recreation: A Household Production Function Approach," Journal of Environmental Economics and Management vol. 5, no. 1 (March) pp. 63-80.

Hay, Michael J. and Kenneth E. McConnell. 1981. "An Analysis of Participation in Nonconsumptive Wildlife Recreation: Reply," Land Economics vol. 57, no. 2 (May) pp. 288-292.

Kmenta, Jan. 1971. Elements of Econometrics (New York, McMillan) 655 pp.

Mallar, Charles D. 1977. "The Estimation of Simultaneous Probability Models," Econometrica vol. 45, no. 7 (October) pp. 1717-1722.

Manski, Charles F. and Steven R. Lerman. 1977. "The Estimation of Choice Probabilities from Choice Based Samples," Econometrica vol. 45, no. 8 (November) pp. 1977-1991.

Pindyck, Robert S. and Daniel L. Rubinfeld. 1976. Econometric Models and Economic Forecasts (New York, McGraw-Hill).

U.S. Department of the Interior, Fish and Wildlife Service. 1968. National Survey of Needs for Hatchery Fish. Bureau of Sport Fisheries and Wildlife, Publication No. 63 (Washington, D.C).

U.S. Department of the Interior. n.d. 1975 National Survey of Hunting, Fishing, and Wildlife-Associated Recreation (NSHFWR) Data tapes and unpublished mimeo, state and national reports (Washington, D.C.).

Chapter 5

VALUING A FISHING DAY

The preceding chapters have described the methods used and the results obtained for the translation of requirements under the Federal Water Pollution Control Act, as amended and later renamed (FWPCA(A)/CWA), into effects on angling participation. This participation was defined in terms of days of fishing with particular classes of fish as major quarry; cold or warm-water game fish and rough fish. The next step in building a national benefit estimate is the determination of values to attach to these days -- that is, individuals' average willingness to pay for days of fishing. In addition, our allocation of water bodies to fishery types, described in chapter 2, rests in part on the assumption of a widely shared preference ranking of classes of fish, and valuation of days spent fishing for these classes will serve as a check on that assumption.

AN OVERVIEW OF ALTERNATIVE APPROACHES TO VALUATION

There are, in principle, two ways to approach the problem of valuing a day's recreational participation at public sites where no entrance or user fees are charged. The first involves the attempt to estimate a representative demand function for the activity. The second is the hypothetical willingness to pay technique, exemplified in the U.S. Fish and Wildlife Service's 1975 National Survey of Hunting, Fishing and Wildlife-Associated Recreation (NSHFWR).

In the direct mail component of NSHFWR, individuals who expressed a preference for fishing over hunting or other kinds of outdoor recreation were asked about their favorite fishing activity: "Having thought about how much this activity cost you in 1975, <u>how much more</u> money would you spend annually on your <u>very</u> favorite activity before deciding to stop doing it because it is too expensive?" The individual's response to this question provides an estimate of his total willingness to pay for a season's fishing experience. Thus, the answer can be viewed as the area under the individual's demand curve for fishing days between the price that drives the individual's annual participation to zero and the shadow price implicit in the actual level of participation he reported (which, at least in theory, is equal to the marginal cost of producing the fishing day (Deyak and Smith, 1978). The quotient of the total willingness to pay and the actual fishing days reported yields the individual's average consumer's surplus (willingness to pay) per fishing day evaluated at the current level of participation. Average daily values computed by Charbonneau and Hay (1978) from the NSHFWR data show that some species are indeed more highly valued than others, and that (except for the anomalous pike case) the value rankings are consistent with those discussed in chapter 2.[1]

Favorite Freshwater Species	Average Consumer Surplus Per Day ($)	Sample Size	Standard Error
Pike, walleye	31	334	12
Trout, landlocked salmon	21	1305	2
Bass	19	853	2
Panfish	19	766	5
Catfish	15	301	3

(From Charbonneau and Hay, 1978)

Demand function estimation, the other broad line of attack on the valuation problem, can itself be subdivided into two approaches, depending upon whether or not one has observed "product" prices which exhibit enough variation to make econometric estimation feasible. Since this is almost never the case for public fishing sites, economists in the recreation field have traditionally relied on the indirect travel cost method to estimate the demand curve for a site's services. (For a pioneering example, see Clawson and Knetsch, 1966.)

Another approach to demand function estimation which presumably would avoid the potential biases of the indirect methods (Bishop and Heberlein, 1979) would be to treat fishing days just like any other marketed commodity, and estimate a demand function from cross-sectional data thrown up by an existing market. There is, in fact, a private market for recreational fishing in the United States, and from data in USDI (n.d.) we calculated that 17 percent of all days spent freshwater fishing in the

United States in 1975 involved payment of a fee for access or use. The work reported in this chapter involves an attempt originally aimed at gathering data to estimate the demand function for recreational fishing at fee sites. Unfortunately, accurate supply-side information necessary to identify the demand function could not be obtained in the survey of operators of fee fishing sites conducted by the authors. Therefore, the approach employed in this chapter involves the estimation of a varying-parameter travel cost model from data on visitation, visitor origins, and site characteristics that we were successful in obtaining (Vaughan, Russell, and Carson, 1981). This refinement of the traditional travel cost model attempts to account for the influence of site characteristics -- especially the fish species available for capture -- on the demand function for the site's services and hence on the average willingness to pay for a day of freshwater recreational fishing differentiated by fish species sought.

The sections that follow report on the estimation of a travel cost demand model from the data collected on a cross-section of fee fishing sites in order to produce average daily value measures for cold-water game fish and rough fish angling.[2] Methodology, data, and results are discussed in turn. In the methodology section we sketch the travel cost method (Clawson and Knetsch, 1966) and its extension to the case of multiple sites with differing characteristics within the context of a varying parameter model (Maddala, 1977). The data section briefly describes our survey sample and touches on some particular problems involving the use of principal components measures of site characteristics, the incorporation of the time costs of travel, and the definition of the model's spatial limits. The results section compares the average values of a fishing day (distinguished by species sought), produced by our variant of the travel cost method with the hypothetical valuation responses from the NSHFWR set out above.

METHODOLOGY: THE VARYING PARAMETER TRAVEL COST MODEL

The travel cost model is based on the insight that frequencies of visitation to a site in response to changes in entry price are similar to the visitation frequencies observed across different levels of travel cost incurred by individuals in moving between their places of residence and the recreational site. By manipulation of the expenditure function, Bowes and Loomis (1980) show that there is an exact theoretical relationship between the consumer's surplus produced by the travel cost method and the consumer's surplus defined as the area under the Hicks-compensated demand function for the site's services. Specifically, except for the unusual case in which an individual would travel to a site but not enter, the consumer's surplus measured by the travel cost method is theoretically equivalent to that measured using entry prices.

We can look at decisions about day-trip excursions to a fishing site as functions of two distinct sets of variables: (1) travel cost and socioeconomic characteristics of the visiting population and (2) attractiveness of the site.

When dealing with grouped data on the visiting population rather than individual interviews, one must assume that all visitors to any particular site in the sample perceive its characteristics similarly. If we were to estimate a single-site travel cost model for any particular site in the cross-sectional sample, we could ignore site characteristics and estimate the parameters of the relationship between visitors per season per zone and zonal population, per capita income, and one-way travel costs. However, to econometrically model the typical visitation-travel cost relationship across several heterogeneous sites, we must account for variation in site characteristics (Ravenscraft and Dwyer, 1978).

This way of looking at the problem of estimating a travel cost model from a multisite sample suggests that the visitation functions can be estimated in two steps (Freeman, 1979). The first step amounts to estimating a separate visitation-travel cost equation for each of the sites in the sample and the second step involves estimating the effect that site characteristics have on the parameters of the equation obtained in the first step. This procedure can be collapsed into one step in which visitation is estimated as a function of travel cost, zonal population, zonal income, <u>and</u> site characteristics, from a pooled sample of all available observations. The algebra demonstrating the equivalence between the two-step and one-step varying parameter models highlights the potential for heteroskedasticity in the error term (Saxonhouse, 1977; Granger and coauthors, 1979).

To obtain an expression for the one-step approach that is functionally equivalent to the two-step approach, substitute (2), below, expressing the first step parameter estimates as functions of site characteristics into (1), the expression for visitation at any particular site and zone pair, iq, to produce (3):

Two-Step Model

$$V_{iq} = \beta_{i0} + \beta_{i1} X_{iq1} + \cdots + \beta_{iK} X_{iqK} + \varepsilon_{iq} \quad (1)$$

$$\begin{aligned}
\beta_{i0} &= \gamma_{00} + \gamma_{01}Z_{i1} + \cdots + \gamma_{0L}Z_{iL} + \eta_{i0} \quad (2)\\
\beta_{i1} &= \gamma_{10} + \gamma_{11}Z_{i1} + \cdots + \gamma_{1L}Z_{iL} + \eta_{i1}\\
&\vdots\\
\beta_{iK} &= \gamma_{K0} + \gamma_{K1}Z_{i1} + \cdots + \gamma_{KL}Z_{iL} + \eta_{iK}
\end{aligned}$$

Equivalent One-Step Model

$$\begin{aligned}
V_{iq} = {} & \gamma_{00} + \gamma_{01}Z_{i1} + \cdots + \gamma_{0L}Z_{iL} \quad (3)\\
& + \gamma_{10}X_{iq1} + \gamma_{11}Z_{i1}X_{iq1} + \cdots + \gamma_{1L}Z_{iL}X_{iq1} + \cdots\\
& + \gamma_{K0}X_{iqK} + \gamma_{K1}Z_{i1}X_{iqK} + \cdots + \gamma_{KL}Z_{iL}X_{iqK}\\
& + \varepsilon_{iq} + \eta_{i0} + \eta_{i1}X_{iq1} + \cdots + \eta_{iK}X_{iqK}
\end{aligned}$$

where

V_{iq} = visitors per year to site i from travel zone q; i=1,...,I and q=1,...,Q

X_{iqk} = first-step explanatory variables for site i zone q including travel costs, per capita income and population; k=1,...,K

$Z_{i(q)\ell}$ = second-step explanatory variables including various measures of site characteristics which are invariant across zones at a given site so $Z_{i(q)\ell} = Z_{i\ell}$ and ℓ=1,...,L

β_{ik} = first-step parameters to be estimated which vary across sites

$\gamma_{k\ell}$ = second-step parameters to be estimated which are invariant across sites

$\varepsilon_{iq}, \eta_{ik}$ = stochastic disturbances

The formulation above shows that the full disturbance term in the one-step model is a function of the explanatory variables in the first stage travel cost regression (travel cost, population, and income per capita), if one believes that every parameter in first stage equations is influenced by site characteristics.

Heteroskedasticity will be absent in the model only if we assume that all first stage parameters except the intercept are <u>not</u> functions of site attributes or that the η's have zero variance. Neither of these assumptions is especially plausible, so a generalized least-squares procedure is called for in estimating such a model to produce unbiased variances of the estimated parameters.

MODEL SPECIFICATION AND DATA COLLECTION

Our cross-sectional data for 1979 were produced from a mail survey of fee fishing sites in the United States conducted by the authors. (For details, see Vaughan, Russell, and Carson, 1981.) At these places, some of which are business ventures and some government operations, individuals either pay a fee per day for the opportunity to fish; a fee per unit of catch (per fish, per inch, or per pound) for fish actually kept; or both. The survey instrument was designed with model specification explicitly in mind.

<u>The Determinants of Demand</u>

Annual visitation at any fishing site is influenced by travel cost (plus the fishing fee paid), the socioeconomic characteristics of the population, the characteristics of the site, and the cost of substitute sites.

The simplest version of the travel cost model includes only the direct variable money cost of making the trip (cost per mile times distance) in the explanatory travel cost variable. This approach, by ignoring the opportunity cost of time spent in travel (Cesario, 1976), produces potentially biased estimates of demand and consumer's surplus unless the marginal utility of travel time is zero and the time spent on site is independent of distance (Wilman, 1980). Individual interview data are required to exactly introduce these further considerations into the travel cost model, since the shadow prices of time spent on site and in transit may differ in magnitude and sign across individuals. Even then, it may be impossible to properly "value" time at the margin (Steiner, 1968).

Our survey data is on sites, not individuals, so there is no neat way out of this dilemma. Therefore we prefer to illustrate the extreme values consumer's surplus can take when the travel cost model is estimated without including the opportunity cost of travel time (TC1), and when an upper limit to travel cost is defined as out-of-pocket cost plus travel time valued at the average wage rate (TC2). Average consumer's surplus derived from a function using TC2 as a regressor will exceed that produced from a function estimated using TC1.

The characteristics of the user population encompass the usual socioeconomic variables so familiar in demand analysis: per capita income, age, sex, education, and the like. It is also useful to think in terms of the aggregate of individuals as defining the site's market. The size of the potential market, while difficult to define precisely, can in general be proxied by the population living within a certain radius of the site. The nature of that market involves in addition the strength of competition faced by any particular site operator from other fee sites and from public sites offering free recreational fishing. Although some of these variables are difficult to measure and the direction of their influence on the demand for fishing days is unclear, they present fewer problems than the variables that capture site characteristics.

The variables that reflect the attractiveness of the sites are difficult to define, and no theoretical model can direct us to a complete and detailed specification. If all sites were the same, the quality differential problem would disappear, but the sites we surveyed differ in a number of ways that we expect are important to anglers.

A central hypothesis of this chapter is that anglers value some species more highly than others, so we expect that the major species class supplied by the fishery is an important site attribute. Our search for other variables qualitatively differentiating the fishing experience led us to several sources outside the normal realm of economics (primarily surveys of angler attitudes) which help shed some light on recreational fishing motivations and the desirable characteristics of the experience. For example, Moeller and Engelken (1972) found in personal interviews that characteristics of the natural environment such as water quality, natural beauty, and privacy while fishing are said by anglers to be more important than the size or number of fish caught. Similarly, Hampton and Lackey (1976) found that for Virginia fisherman at fee sites, aesthetic factors were claimed to be more important than the size or number of fish caught. Further, in a study of the motives of salmon anglers in British Columbia, Brian found that being outdoors and discharging tension were more frequently mentioned motives for fishing trips than the experience of a catch (SFI, 1974). Studies of Rhode Island boat-using sport fishermen also suggest that motives other than catch are important to fishermen (Spaulding, 1971).

These general observations, the arguments of Wennergren and coauthors (1975), and the empirical work of Holman and Bennett (1973) lead us to a more specific list of site characteristic variables which should be measured in any cross-sectional site survey designed to gather data on the demand side:

1. Average number of fish caught per visitor day
2. Average weight of fish caught per visitor day
3. Species mix of the expected bag
4. Type of water body (flowing or standing water)
5. Quality of aesthetic background
6. Provision of complementary facilities including: boat rental, equipment rental, sanitary facilities
7. Accessibility of site - proximity to a major highway
8. Weather - number of "fishable" days per year
9. Fishable acreage of site
10. Site congestion
11. Rules and regulations - bag, equipment, or time limits, catch and keep rules
12. Extent of competition experienced from substitute private and public sites

Data on all these prospective explanatory variables were collected in our fee fishery survey. The survey data were supplemented with data on population size, characteristics, and climate (indexed by site and zone) produced from HEW's Area Resource File using a computer program designed for that purpose.

DEFINITION OF THE DEPENDENT VARIABLE

The source of our information about visitation was the site owner-operator, who was asked to report the total number of visitors to his site for the 1979 season, and their distribution across seven origin zones (0 to 10 miles; 10 to 25 miles; 25 to 50 miles; 50 to 100 miles; 100 to 150 miles; 150 to 200 miles; and over 200 miles). The visitors (if any) from the last, open-ended zone were not included in the data set on the rationale that the assumption that all observed trips are taken for the primary purpose of visiting the site to fish probably does not hold beyond a 200-mile radius (Smith and Kopp, 1980).[3]

A theoretical maximum market radius of 200 miles for all sites does not mean that 200 miles was the effective market radius actually observed for any fee site. Some sites had visitors from over 200 miles, and others showed zero visitation from several of the zones within the 200-mile boundary. One question of some interest is whether or not to include as data every zero visitation rate observed for every site.

If we regard the statistically estimated visitation-travel cost relationship as a representation of the demand curve for a site's services by individuals residing in a particular zone (Wilman, 1980, p. 279) it is defined only in the northeast quadrant of price-quantity space. The representative recreationist's reservation price is the price driving quantity demanded to zero (Varian, 1978, p. 104). This is defined by the data point describing the "effective first zero" which represents the travel cost at which visitation drops to zero. Travel costs higher than this reservation price which are also associated with zero visitation are irrelevant unless they can be explained by some taste variables in the first stage regression. Our data on income and taste variables are inadequate for this purpose, so such zero zones were not included in the data set.

We included zonal population as an explanatory variable in an equation estimating visitors per season as a function of travel cost and other exogenous variables, rather than deflating the dependent variable, visitors, by zonal population to estimate a relation between travel cost and visitation per thousand population. Although the visits-per-capita form is far more common, there is some precedent for the visitors form (Merewitz, 1966; NERA, 1974). The visits-per-capita form can result in misprediction of visits when the estimated equation is evaluated at the means, and can be heteroskedastic. Heteroskedasticity could also be present in the trips model either because of grouping of observations over distance zones or because of the heteroskedasticity potentially present in the varying parameter model. Therefore a constructive test for heteroskedasticity is undertaken in the estimation step discussed below.

A list of the variables in the model appears in table 5-1. A descriptive analysis of the site survey data appears in Vaughan and coauthors (1981) so for the sake of brevity the means of the variables are not repeated here.

In the pooled one-step equation, the variables for population and site characteristics must appear separately along with the interaction of each population characteristic with each site characteristic (eq. 3). In general, we have no a priori expectations about the signs of the coefficients of many of these site characteristics or of their interactions

Table 5-1. Variables in the Site Survey Sample

Variable	Description	Source
I. Dependent		
VZONE	Visitors to site indexed by zone of origin	RFF Survey
II. Independent: Population Characteristics		
TC1 or TC2	One-way zone-to-site travel cost plus fishing fee (TC1) or one-way travel cost plus fishing fee plus opportunity cost of travel time valued at hourly wage rate per zone (TC2)	o AAA for variable cost o RFF Survey for fishing fee o BEA for average hourly wage rate (TC2 only)
CAPINC	Average zonal per capita income ($ 1975)	ARF
POPULZ	Zonal population (000)	ARF
III. Independent: Site Characteristics		
MILE4	Distance of the site from a major four-lane highway (in miles); an indicator of accessibility	RFF Survey
FLOWING	Dummy variable to indicate the existence of a river or stream that can be fished from (1) or not (0)	RFF Survey
TOTACRE	Total surface acreage of all lakes and ponds at the site	RFF Survey
USEBANK	Percentage of the circumference of the bank around all lakes and ponds at the site which can be fished from	RFF Survey
EQRNT	Dummy variable to indicate the availability of equipment rental (1) or not (0)	RFF Survey
FOOD	Dummy variable to indicate the availability of food from a restaurant or concession stand (1) or not (0)	RFF Survey

Table 5-1 continued

ONIGHT	Dummy variable to indicate the availability of overnight lodging or camping facilities (1) or not (0)	RFF Survey
PIC	Dummy variable to indicate the availability of picnic facilities (1) or not (0)	RFF Survey
SHOW	Dummy variable to indicate the availability of showers (1) or not (0)	RFF Survey
ANGLIM	Dummy variable to indicate whether the site operator restricts the number of anglers allowed to fish on a given day (1) or not (0)	RFF Survey
BOATS	Dummy variable to indicate whether anglers are allowed to fish from boats (1) or not (2)	RFF Survey
CATCHUM	Dummy variable to indicate whether there is a catch limit imposed (1) or not (0)	RFF Survey
PERTHRO	Percent of fish caught which are thrown back	RFF Survey
ADULTPER	Percent of total visitors who are adults	RFF Survey
SEASONDA	Total number of days the site is open per year	RFF Survey
SCAPE	Dummy variable to indicate whether landscaping has been done (1) or not (0)	RFF Survey
LOSS	Dummy variable to indicate whether the site is losing money (1) or not (0)	RFF Survey
FCOMALOT, FCOMSOME, FCOMLIT	Dummy variables to indicate whether the site is exposed to a lot, some, or a little competition from fee sites in the market area	RFF Survey
PCOMALOT PCOMSOME PCOMALIT	Dummy variables to indicate whether the site is exposed to a lot, some, or a little competition from public sites in the market area	RFF Survey

Table 5-1 continued

HILL	Dummy variable to indicate whether the terrain surrounding the site is either hilly/mountainous (1) or flat (0)	RFF Survey
WOODS	Dummy variable to indicate whether the site is surrounded by woods (1) or not (0)	RFF Survey
REPRO	Dummy variable to indicate whether there is moderate-to-excellent natural reproduction (1) or poor-to-no natural reproduction (0) in the fishery	RFF Survey
CRDBNKDA	Dummy variable to indicate whether there is crowding among fishermen on the bank during weekdays (1) or not (0)	RFF Survey
CRDBNKEN	Dummy variable to indicate whether there is crowding among fishermen on the bank during weekends (1) or not (0)	RFF Survey
JATEMP	Mean January Temperature at site ^{o}F	ARF
JLTEMP	Mean July temperature at site ^{o}F	ARF
RAIN	Mean annual precipitation at site in inches	ARF
TBAGNO	Average number of trout caught per fisherman day	RFF Survey
BBAGNO	Average number of bass caught per fisherman day	RFF Survey
PBAGNO	Average number of panfish caught per fisherman day	RFF Survey
CBAGNO	Average number of catfish and carp caught per fishermen day	RFF Survey
TWGHT	Weight of the average trout caught (pounds)	RFF Survey
BWGHT	Weight of the average bass caught (pounds)	RFF Survey
PWGHT	Weight of the average panfish caught (pounds)	RFF Survey

Table 5-1 continued

CWGHT	Weight of the average catfish-carp caught (pounds)	RFF Survey

Sources:

AAA: American Automobile Association, "Your Driving Costs" (Falls Church, Virginia, 1980)

ARF: U.S. Department of Health, Education and Welfare, Bureau of Health Manpower, Area Resource File (Data Tape)

BEA: U.S. Department of Commerce, Bureau of Economic Analysis, Local Area Personal Income, 1973-1978 (Data Tape)

RFF Survey: Resources for the Future, Recreational Fee Fisheries Survey (1979) with the first stage variables. Nor are we interested in the particular effect of any of these characteristic measures. Rather, our purpose is to control for their effects in order to produce a demand schedule for fishing days at a site of "typical" or average quality in its species class. We therefore reduce the dimensionality of the problem and "substantially" account for variations in the site characteristics variables by constructing principal components (Dhrymes, 1970).[4] The scaled principal components scores (zero mean, unit variance) for each site, computed from the standarized values of the site characteristics variables, are employed as independent variables in the visitation-travel cost regressions, leaving travel cost, population, and income in their original units of measurement. (For examples of this approach, see Kopp,1980; Smith and Deyak, 1975.)

RESULTS: MODEL ESTIMATION AND EVALUATION

The fish species supplied to anglers is the characteristic variable of critical interest in this study. Fortunately, the fee fishing sites we surveyed tend to specialize in one species or another. By assigning a site to a particular fishery type if 70 percent or more of the weight of fish caught was from that species group we produced 83 usable returns in the trout class and 66 in the catfish/rough fish class.[5] The full model described above was fit separately to the data in each of these subsamples. For each subsample distinguished by species we estimated a pair of equations, the difference in each case being the travel cost variable used: TC1 (no allowance for the cost of time spent traveling) and TC2 (with time spent traveling valued at the median hourly wage by travel zone). All equations were estimated using Park's constructive pseudo-GLS procedure (discussed in appendix D), since mild heteroskedasticity was detected in every case (Park, 1966).[6] The treatment of zero zonal visitation observations was to eliminate all but the first effective (noninterstitial) zero.

The results of the regression analysis appear in table 5-2, where for compactness the sixty-odd parameter estimates for the principal components scores and their interactions with the first stage variables are not shown.

The coefficients of determination indicate that all equations in each model fit the data well.[7] The estimated coefficients on the variants of travel cost (TC1, TC2) are significant at the 5 percent level for both the

Table 5-2. One-Step Varying Parameter Travel Cost Models for Trout and Catfish Sites

(absolute values of "t" statistics in parentheses)

Variable	Trout: Travel cost only	Trout: Travel and time cost	Catfish: Travel cost only	Catfish: Travel and time cost
INTERCEPT	367.08 (1.71)	372.14 (1.82)	242.50 (0.50)	228.27 (0.50)
TC1	-14.73 (2.23)*	... ...	-38.22 (2.80)**	... ...
TC2	... ...	-8.53 (2.86)**	... ...	-20.35 (2.93)**
CAPINC	0.036 (0.66)	0.021 (0.40)	0.179 (1.61)	0.149 (1.38)
POPULZ	0.030 (1.86)	0.035 (2.28)*	-0.079 (2.23)*	-0.043 (1.04)
R^2	0.38	0.41	0.50	0.53
F	3.90**	4.46**	4.29**	4.84**
Number of observations	471	471	316	316
Number of parameters	64	64	60	60

Notes:

*Indicates significance at 5% level.

**Indicates significance at 1% level.

trout and catfish/rough fish regressions. Collinearity did not appear to degrade the travel cost coefficient, but may be the cause of one coefficient -- that attached to population in the catfish regression -- having an unexpected negative sign.[8]

To test for whether or not site characteristics influence the step-one parameters (and hence price and income elasticities of demand) we posited a reduced model. Its arguments were the step-one variables and the principal components on site characteristics but not the interaction terms between site characteristics and step-one variables. The null hypothesis is that the parameters attached to the interaction terms are zero. On the basis of the F tests we reject the null hypothesis and conclude that the slope parameters of the step-one travel cost model are influenced by site characteristics.

The equations in table 5-2 represent zonal demand functions. To produce an aggregate demand curve for the services of a representative trout or catfish site in the sample by a representative group of individuals, several additional steps must be performed (summarized in Freeman, 1979, p. 203). First, zonal means of the independent variables in the sample must be calculated. Second, predicted visitation from each zone, $VHAT_q$, under the existing travel cost and fee structure must be obtained from the estimated equation using the zonal means of the variables. Then given $VHAT_q$, the individual zonal demand curve expressed as a function of price is:

$$V_q = VHAT_q - \hat{\beta}_1 P + \sum_{i=1}^{K} (\hat{\beta}_{i+1} (P \cdot PRINCOMP_i))$$

where β_1 is the estimated coefficient in travel cost, β_{i+1} is the estimated coefficient on the appropriate travel-cost-component interaction term, P is a hypothetical increment or decrement in entry price from the reigning fee such that $V_q \geq 0$, and PRINCOMP is the component score. Since the sample mean of each principal component score is zero (Massy, 1965), the aggregate demand curve across all possible prices evaluated at the means of all other variables is traced out as sum of the zonal visitation levels across alternative values of P:

$$V = \sum_{q=1}^{Q} (VHAT_q - \hat{\beta}_1 P)$$

The demand schedules described in this manner are shown in figures 5-1 and 5-2 for typical trout and catfish sites respectively.

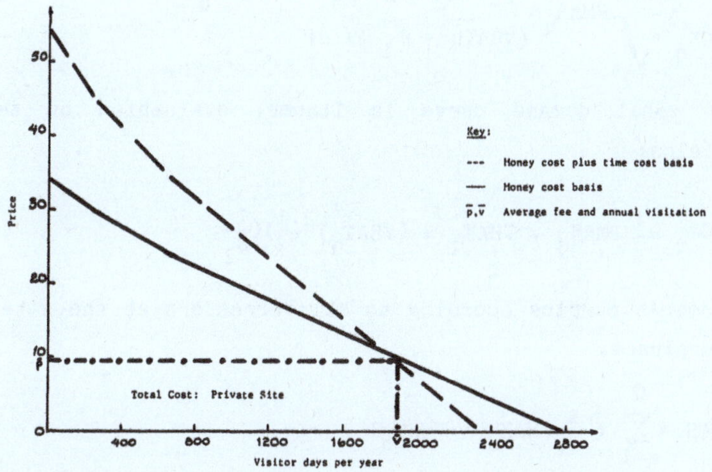

Figure 5-1. Typical fee site demand schedule: Trout sample.

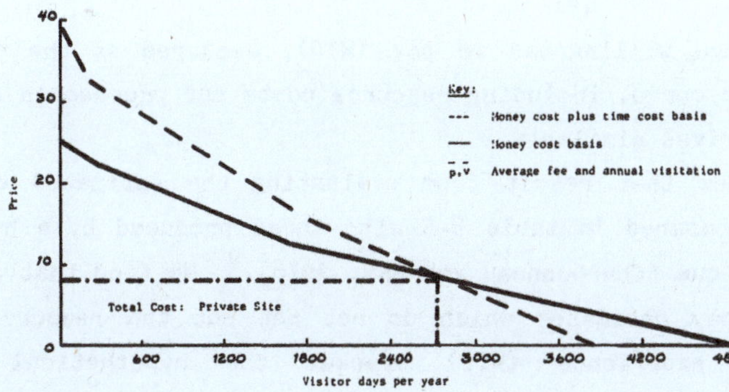

Figure 5-2. Typical fee site demand schedule: Catfish/roughfish sample.

The upper limit of integration for any zonal demand curve is the price $PMAX_q$ driving zonal visitation to zero:

$$PMAX_q = VHAT_q / \hat{\beta}_1$$

Total zonal consumer's surplus is the area under any zonal demand curve from zero (no increment to existing fee) to $PMAX_q$:

$$CS_q = \int_0^{PMAX_q} (VHATq - \hat{\beta}_1 P)\, dP$$

Since each zonal demand curve is linear, evaluation of the definite integral yields:

$$CS_q = \frac{1}{2} PMAX_q \cdot VHAT_q = (VHAT_q)^2 (\tfrac{1}{2})(\tfrac{1}{\beta_1})$$

Total consumer's surplus accruing to all recreators at the site is the sum of zonal surpluses:

$$CS = \sum_{q=1}^{Q} \left(\frac{1}{2} PMAX_q \cdot VHAT_q \right)$$

Average consumer's surplus per day is:

$$\overline{CS} = \sum_{q=1}^{Q} \frac{1}{2} \frac{(PMAX_q \cdot VHAT_q)}{\sum_{q=1}^{Q} VHAT_q}$$

Total and average willingness to pay (WTP), measured as the entire area under the demand curve, including resource costs and producer's surplus (if any), can be derived similarly.

The surpluses that result from evaluating the estimated travel cost equations are compared in table 5-3 with those produced by a hypothetical valuation technique (Charbonneau and Hay, 1978).[9] We find that our average willingness to pay estimates which do not net out the resource costs of supplying the experience (WTP) bracket the hypothetical valuation estimates. These results are encouraging for our maintained hypothesis of the differential valuation of a day's fishing according to species sought. This ranges between $4 and $8 per day difference in WTP between trout and catfish angling.

Unfortunately, we cannot produce an average daily value for warm-water game fish/panfish angling because of the limitations of our site data. However, we can use the Charbonneau and Hay, (1978) information on the value of a bass-fishing day to interpolate. This procedure yields the average surpluses employed in the next chapter:

Table 5-3. Comparison of Average Values of a Fishing Day by Species Sought

	Average surpluses	
	Excluding resource costs ($\overline{CS}$)	Including resource costs ($\overline{WTP}$)
<u>Travel cost method</u>		
Trout		
TC1	10.96	15.60
TC2	19.49	24.09
Catfish		
TC1	7.00	10.62
TC2	12.48	16.03
<u>Hypothetical valuation method</u>[a]		
Trout	...	21.00
Catfish	...	15.00

[a]Computed from responses to the U.S. Department of the Interior's <u>1975 National Survey of Hunting, Fishing and Wildlife Associated Recreation</u> by Charbonneau and Hay (1978).

Type of Angling	Value Per Day Low	High
Cold-water game fish [a]	10.96	24.09
Warm-water game fish/panfish [b]	9.65	21.43
Catfish/rough fish [a]	7.00	16.03

[a]Low figures for cold-water game fish and catfish/rough fish were produced from our travel cost model using TC1 and excluding resource costs from the surplus measures. High figures reflect the use of TC2 and inclusion of resource costs in the surplus measure.

[b]Equal to our catfish value plus two-thirds of the difference between our trout and catfish values. Interpolation factor of two-thirds from Charbonneau and Hay (1978).

CONCLUSIONS

The estimation of a varying parameter travel cost model discussed in this chapter was conducted using less than ideal data, and several difficult econometric problems had to be solved. Given the nature of our data, one problem that could not be addressed directly was the cost of substitute sites (Burt and Brewer, 1971), which we could only proxy by dummy variables reflecting the extent of competition from other sites perceived by the site owner.

However, the method discussed here represents a simple, straightforward way of incorporating site characteristics into the travel cost framework. Its application permits the valuation of changes in site attributes, such as fish species supplied or levels of bag provided by management measures such as stocking. When the participation impacts of a broad public policy such as water pollution control must be valued, the method is especially useful if quantities and values must be distinguished in some quality dimension of the experience. In our application, the model produced average daily willingnesses to pay distinguished by fish species sought (inclusive of resource costs) consistent with values obtained from a hypothetical valuation experiment.

We now have generated all the building blocks required for the estimation of the national benefits of existing water pollution control as they accrue to freshwater fishermen. The next chapter combines the blocks and analyzes the resulting benefit numbers.

NOTES

1. This does not imply that such rankings are universal across all regions of the United States. There are perhaps individuals who would value a day's catfishing more highly than a day's trout fishing, even if bag and all other variables were held constant.

2. Our survey did not turn up enough sites that offered strictly (or at least very largely) warm-water game fishing to allow us to estimate a bass/panfish day value. Because, as we note below, our estimates for trout and rough fish days match the NSHFWR results quite closely, we are able to interpolate to obtain a value for a day of bass and panfishing.

3. From McConnell (1975) we know that the theoretically correct unit of measurement for the dependent variable in the travel cost approach should be trips (or visits), not recreation days (or visitors). Generally, our sites appeal to individuals planning a day trip, not an extended vacation. By limiting the market radius to 200 miles or less and by deleting from the sample sites which are primarily for overnight camping, we have attempted to ensure that no trips are multiday excursions, so visitors, recreation days, and trips are nearly equivalent in our sample.

4. Principal component analysis of thirty-nine site characteristics in the trout sample produced fifteen factors which explain 74 percent of the variance in the site characteristics. For the rough fish sample, fourteen components were produced from forty site characteristics and explain 73 percent of their variances. This procedure may introduce some bias into the parameter estimates on the original variables, influence their computed variances, and make the validity of the reported "t" statistics questionable. It was undertaken solely to reduce the dimensionality of the problem in the face of limited survey data.

5. See above, note 2, where we explain that our sample of warm-water game fishing sites was too small to allow estimation.

6. The Park test is a special case of Harvey's (1976) test for multiplicative heteroskedasticity. Harvey argues that the multiplicative heteroskedasticity model is more attractive than the additive model, which assumes that the variance of the ith disturbance terms is a linear combination of known variables, as in Eq. 3. in this chapter. Harvey's likelihood ratio test of the null hypothesis of homoskedastic errors is asymptotically equivalent to the F test employed in appendix D (Judge and coauthors, 1982, chapter 14). The weights matrices used to estimate all pseudo-GLS models discussed in this section were generated using all three estimated slope parameters, δ_i, from the Park regressions of the log of the OLS residuals on the logs of travel cost (TC), per capita income (CAPINC) and population (POPULZ):

$$WT_i = \frac{1}{TC_i^{\hat{\delta}} \cdot CAPINC_i^{\hat{\delta}_2} \cdot POPULZ_i^{\hat{\delta}_3}}$$

7. The data analysis program we used to estimate these equations, SAS, produces a weighted R^2 measure when weighted regression is performed with a specified weighting matrix. The formula is:

$$R^2 \text{ SAS, weighted} = 1 - \frac{\sum \hat{e}_i^2 \cdot WT_i}{\sum WT_i \left(Y_i - \bar{Y}_w\right)^2}$$

where

WT_i = i^{th} element on the main diagonal of the Ω^{-1} weighting matrix

$\bar{Y}_W$ = weighted mean of the dependent variable:

$$\bar{Y}_W = \frac{\sum Y_i \cdot WT_i}{\sum WT_i}$$

$\hat{e}_i$ = unweighted residuals from the weighted regression using its parameter estimates, $\hat{\beta}$, and the untransformed dependent and independent variables:

$$\hat{e}_i = Y - X\hat{\beta}$$

For a discussion of formulas for R^2 in weighted regression, see Blomquist, 1980.

8. Singular-value decomposition (SVD) of the scaled X matrix (Belsley, Kuh, and Welsch, 1980) and auxiliary regressions revealed moderate to strong pairwise near dependencies among the intercept and per capita income, and also between each of the principal component scores and its interactions with per capita income. This situation arose because per capita income is nearly constant across zones and sites in our data. This was true for both the trout sites and catfish sites examined separately. The fortunate and important observation for our purposes is that the estimated travel cost parameter was found to be uncontaminated by collinearity. This key parameter is "buffered" because its variance does not have a large proportion of its magnitude associated with the presence of the moderate to strong collinearity found elsewhere.

9. The responses to the hypothetical valuation questions in the U.S. Department of the Interior's National Survey of Hunting, Fishing and Wildlife (NSHFWR) are conceptually closest to a travel cost valuation measure which includes the opportunity cost of travel time and the resource cost of providing the experience.

Although the NSHFWR data are in 1975 dollars, and our travel cost measure, obtained from the AAA (table 5-1), is in 1979 dollars (7.62 cents per mile), the latter seems on the low side compared with earlier studies which, even before the era of high fuel prices, often used 10 cents per mile. It is hard to believe car owners substituted away from low mileage vehicles in the face of rising fuel prices to such an extent that the variable cost per mile of operating a vehicle fell in the 1970s. Therefore, we have not converted the NSHFWR data summarized in Charbonneau and Hay into 1979 dollars.

REFERENCES

Belsley, David A., Edwin Kuh, and Roy E. Welsch. 1980. Regression Diagnostics: Identifying Influential Data and Sources of Collinearity (New York, Wiley).

Bishop, Richard C. and Thomas A. Heberlein. 1979. "Measuring Values of Extramarket Goods: Are Indirect Measures Biased?" American Journal of Agricultural Economics vol. 61, no. 5 (December) pp. 926-930.

Blomquist, N. Soren. 1980. "A Note on the Use of the Coefficient of Determination," Scandinavian Journal of Economics vol. 82, no. 3, pp. 409-412.

Bowes, Michael D. and John B. Loomis. 1980. "A Note on the Use of Travel Cost Models with Unequal Zonal Populations," Land Economics vol. 56 (November) pp. 465-470.

Burt, Oscar R. and Durward Brewer. 1971. "Estimation of Net Social Benefits from Outdoor Recreation," Econometrica vol. 39, no. 5 (September) pp. 813-827.

Cesario, Frank J. 1976. "Value of Time in Recreation Benefit Studies," Land Economics vol. 52 (February) pp. 32-41.

Charbonneau, John and Michael J. Hay. 1978. "Determinants and Economic Values of Hunting and Fishing." A paper presented at the 43rd North American Wildlife and Natural Resources Conference, Phoenix, Arizona, March 18-22, 20 pp.

Clawson, Marion, and Jack L. Knetsch. 1966. Economics of Outdoor Recreation (Baltimore, Md., Johns Hopkins University Press for Resources for the Future).

Deyak, Timothy A. and V. Kerry Smith. 1978. "Congestion and Participation in Outdoor Recreation: A Household Production Function Approach," Journal of Environmental Economics and Management vol. 5, no. 1 (March) pp. 63-80.

Dhrymes, Phoebus J. 1970. Econometrics: Statistical Foundations and Applications (New York, Springer Verlag).

Freeman, A. M. III. 1979. The Benefits of Environmental Improvement (Baltimore, Md., Johns Hopkins University Press for Resources for the Future).

Granger, Clive W.J., Robert Engle, Ramu Ramanathan, and Alan Anderson. 1979. "Residential Load Curves and Time-of-Day Pricing," Journal of Econometrics vol. 9, pp. 13-32.

Hampton, E. L. and R. T. Lackey. 1976. "Analysis of Angler Preferences and Fisheries Management Objectives with Implications for Management," Proceedings of the Southeastern Association of Game and Fish Commissioners vol. 29, pp. 310-316.

Harvey, A. C. 1976. "Estimating Regression Models with Multiplicative Heteroscedasticity," Econometrica vol. 44 (May) pp. 461-465.

Holman, Mary A. and James T. Bennett. 1973. "Determinants of Use of Water-based Recreational Facilities," Water Resources Research vol. 9 (October) pp. 1208-1218.

Judge, George G., R. Carter Hill, William E. Griffiths, Helmut Lutkepohl, and Tsoung-Chao Lee. 1982. Introduction to the Theory and Practice of Econometrics (New York, Wiley).

Kopp, Raymond J. 1980. "Energy, Residuals and Inefficiency: An Engineering Econometric Analysis of Environmental Regulations," in Advances in the Economics of Energy and Resources III, John Moroney, ed., (Greenwich, Conn., JAI Press) pp. 199-219.

Maddala, G. S. 1977. Econometrics (New York, McGraw-Hill).

Massy, William F. 1965. "Principal Components Regression in Exploratory Statistical Research," Journal of the American Statistical Association vol. 60 (March) pp. 234-256.

McConnell, Kenneth E. 1975. "Some Problems in Estimating the Demand for Outdoor Recreation," American Journal of Agricultural Economics vol. 57 (May) pp. 330-334.

Merewitz, Leonard. 1966. "Recreational Benefits of Water Resource Development," Water Resources Research (Fourth Quarter) pp. 625-640.

Moeller, George H. and John H. Engelken. 1972. "What Fishermen Look for in a Fishing Experience," Journal of Wildlife Management vol. 36, no. 4 (October).

National Economic Research Associates (NERA). 1974. "Estimates of the Value of Changes in Aquatic Life Resulting from Reductions in Thermal Effluent." Attachment V to Edison Electric Institute Ad Hoc Water Quality Group and Utility Water Act Group, Comments on EPA's Proposed 304 Guidelines and 306 Standards of Performance for Steam Electric Power Plants. Report submitted to U.S. Environmental Protection Agency, Washington, D.C.

Park, R.E. 1966 "Estimation with Heteroskedastic Error Terms" Econometrica vol. 34, no. 4 (October) p. 888.

Ravenscraft, David J. and John F. Dwyer. 1978. "Reflecting Site Attractiveness in Travel-Cost-Based Models for Recreation Benefit Estimation," Forestry Research Report No. 78-6 (Urbana, Ill., University of Illinois at Urbana-Champaign Agricultural Experiment Station).

Saxonhouse, Gary R. 1977. "Regression from Samples Having Different Characteristics" <u>Review of Economics and Statistics</u> vol. 59 (May) pp. 234-237.

Smith, V. Kerry, and Timothy A. Deyak. 1975. "Measuring the Impact of Air Pollution on Property Values," <u>Journal of Regional Science</u> vol. 15, no. 3, pp. 277-288.

______ and Raymond J. Kopp. 1980. "The Spatial Limits of the Travel Cost Recreational Demand Model," <u>Land Economics</u> vol. 56 (February) pp. 64-72.

Spaulding, Irving A. 1971. "Occupation, Recreation and Psychic Communication: Selected Rhode Island Sport FishermEn," Bulletin No. 405. (Kingston, R.I., University of Rhode Island Agricultural Experiment Station).

Sport Fishing Institute. 1974. <u>SFI Bulletin</u> No. 258 (September).

______, 1977. <u>SFI Bulletin</u> No. 286 (July).

Steiner, Peter O. 1968. "Comment on 'The Value of Travel Time,'" in Samuel B. Chase, ed., <u>Problems in Public Expenditure Analysis</u>, (Washington, D.C., Brookings Institution).

U.S. Department of the Interior. n.d. <u>1975 National Survey of Hunting, Fishing, and Wildlife-Associated Recreation</u> (NSHFWR) Data tapes and unpublished mimeo, state and national reports (Washington, D.C.).

Varian, Hal. R. 1978. <u>Microeconomic Analysis</u> (New York, W. W. Norton).

Vaughan, William J., Clifford S. Russell, and Richard T. Carson. 1981. "A Survey of Recreational Fee Fishing Enterprises," <u>Farm Pond Harvest</u> vol. 16 (Fall) pp. 12-14, 22, 26.

Wennergren, E. Boyd, Herbert H. Fullerton, John E. Keith, and Robin Meale. 1975. "Economic Value of Water-Oriented Recreation Qaulity," Utah Water Research Laboratory Study PRRAE 805-1 (Logan, Utah, Utah State University).

Wilman, Elizabeth A. 1980. "The Value of Time in Recreation Benefit Studies," <u>Journal of Environmental Economics and Management</u> vol. 7 (September) pp. 272-286.

Chapter 6

RESULTS AND CONCLUSIONS

After several chapters of painful econometric detail, the reader may be feeling quite out of sight of the forest. So, before discussing our results we first recapitulate the method by which we pursued estimates of the national benefits of water pollution control accruing via recreational fishing.

On the basis of theoretical arguments for the importance of supply side variables in determining participation in such recreation activities as fishing, we hypothesized that water pollution control would result in more total days of recreational fishing. This would happen both because more individuals out of the population will choose to fish when more fishable water is available and because those people who do fish will on average fish more days per year. In addition, we expected pollution control to result in higher proportions of game fishing water (cold and warm) because of the greater sensitivity of game fish to pollution and the desire of state fish and game managers to increase the availability of game fishing water whenever possible. This, in turn, we hypothesized would lead to relatively greater increases in game fishing. Finally, we argued (and demonstrated) that days of game fishing are, on average, valued more highly than days of rough fishing. Thus we expected to find that pollution control stimulated both more and more highly valued freshwater recreational fishing.

SUMMARY OF KEY RESULTS

Table 6-1 brings together in summary form the results of our actual estimations and projections, beginning on the left with the effects of water pollution control on the extent and composition of fishable water and moving through participation probabilities and days of participation per

Table 6-1. Summary of Results for Participation Changes

Scenario	Sample and statistical variants		Total fishable water (10^3 acres)	Fishery shares			Probability of being a fisherman	Total fishermen (10^6)	Probability of doing some			Days per capita per year			Total days per year of		
	Sample[a]	Method[b]		Trout	Bass	Rough			Trout fishing	Bass fishing	Rough fishing	Trout fishing	Bass fishing	Rough fishing	Trout fishing (10^6)	Bass fishing (10^6)	Rough fishing (10^6)
BASE	5,000	LOGIT	30,615	0.204	0.684	0.112	0.2793	59.16	0.3742	0.6919	0.3287	14.62	18.44	9.89	323.6	754.6	192.3
	11,000	WLS	30,615	0.204	0.684	0.112	0.2793	59.16	0.3708	0.6840	0.3499	13.76	18.22	10.14	301.8	737.4	209.8
BPT	5,000	LOGIT	30,721	0.227	0.736	0.037	0.2794	59.18	0.3818	0.6777	0.3248	14.64	18.78	10.30	330.8	753.3	197.9
	11,000	WLS	30,721	0.227	0.736	0.037	0.2794	59.18	0.3931	0.6776	0.3536	13.73	18.48	10.55	319.3	741.4	220.8
BPT[c] +S	5,000	LOGIT	30,934	0.231	0.742	0.027	0.2796	59.22	0.3832	0.6755	0.3239	14.65	18.83	10.35	332.4	753.3	198.5
	11,000	WLS	30,934	0.231	0.742	0.027	0.2796	59.22	0.3964	0.6763	0.3538	13.72	18.53	10.60	322.2	742.2	222.1
BPT +S pH[d]	5,000	LOGIT	31,134	0.229	0.743	0.028	0.2798	59.26	0.3821	0.6765	0.3247	14.63	18.84	10.34	331.4	755.2	199.0
	11,000	WLS	31,134	0.229	0.743	0.028	0.2798	59.26	0.3954	0.6771	0.3544	13.71	18.54	10.60	321.3	743.8	222.5
"BAT"[e]	5,000	LOGIT	31,247	0.232	0.755	0.013	0.2799	59.28	0.3837	0.6741	0.3244	14.63	18.92	10.43	332.2	755.9	200.5
	11,000	WLS	31,247	0.232	0.755	0.013	0.2799	59.28	0.3992	0.6766	0.3557	13.70	18.60	10.68	324.1	745.9	225.2

[a]Size of sample used in estimating fishery-type participation and days equations.

[b]Method of estimating fishery-type participation equation.

[c]BPT + S is BPT imposed on point sources and assumed control of nonpoint sources of suspended solids loads. (S)

[d]BPT + S + pH is BPT plus suspended solids control plus control of acid mine drainage.

[e]"BAT" is a requirement that all streams have at least 5 ppm dissolved oxygen, unless it is naturally unattainable, at summer low flows.

capita, to total annual days of fishing, by type of fish sought and in total, on the right. These results are presented for the base (precontrol) situation and for the four stages of policy implementation identified in chapter 2. For each policy stage we also show the effects of varying the basis of our estimates; in particular, the character of the sample of survey respondents used in estimating the fishing-type participation and days equations and the method of estimation (logit or WLS) used for the participation probability relations.

In several important respects, the results confirm our expectations. More people become anglers, and each angler is projected, on average, to spend more days per year fishing (with a few exceptions in the trout category). On the other hand, the hypothesis that game fishing would gain in relative importance is not confirmed. While total days of game fishing per year are projected to increase (markedly for trout, much less for bass), rough fishing days grow slightly faster and account for an increasing share of total fishing days. This phenomenon is illustrated in table 6-2 for three scenarios and two combinations of sample size and estimation technique. Thus, we project that while upgrading is going on, and while rough water falls as a percentage of total fishable water, rough fish anglers still increase significantly the time they spend fishing. This result was anticipated by some of the unexpected coefficient sign patterns reported in chapter 4, especially those for rough fishing water's influence on rough fishing days. That negative sign translates a decrease in rough water acreage into an increase in days of fishing per rough fish angler.

Table 6-2. Shares of Total Annual Angling Days Accounted for by Fishery Types

Scenario	Sample size	Estimation method	Percent of total days done: Trout fishing	Bass fishing	Rough fishing
Base	5,000	LOGIT	25.2	59.4	15.1
	11,000	WLS	24.2	59.0	16.8
BPT	5,000	LOGIT	25.8	58.8	15.4
	11,000	WLS	24.9	57.9	17.2
BAT	5,000	LOGIT	25.8	58.7	15.6
	11,000	WLS	25.0	57.6	17.4

It is also true that the participation probabilities do not behave as we expected. The probability that a fisherman will do some bass fishing in a year falls as pollution control proceeds. The same is true, using the 5,000-person sample, for the probability of doing some rough fishing. In general, it seems to us that these problems may be traced to the difficulty of separating bass and rough fishing over much of the water in the United States. That is, warm, clean, natural water bodies will often hold respectable stocks of both bass (and panfish) and rough fish, particularly catfish. We classify these waters as game fishing, but local rough fishermen may see them as God's own gift to their sport. This effect is probably compounded by regional preference differences, though as we mentioned in chapter 4, the introduction of a dummy variable to stand for special tastes in the South failed to change the estimated signs of the coefficients relating water availability to participation probabilities and days of participation.[1]

Despite these unexpected twists, total days of fishing of all kinds are projected to increase by 18 to 45 million days per year as pollution control proceeds from the base case to BAT. The direction of the results is insensitive to choice of sample size and estimation technique for the fishing-type participation equations. But the combination does make some big differences in the size of the predicted effects. The 11,000 sample, using weighted least squares for both participation and days equations seems to us the best information source. It produces the largest participation change predictions.

VALUING THE CHANGES IN PARTICIPATION AND DERIVING POLICY BENEFITS

In table 6-3 we repeat the estimates of the value of fishing days, by type of fish sought, as derived from our fee-fishery data and travel cost model in chapter 5. Again, the variations in these values with certain key assumptions are included. They show how the estimated values shift with assumptions about the value of travel time and the extent to which the fees charged at the sample sites reflect true costs or only special rents (transfers). The "high values" assume that travel time should be valued at the wage rate and that the marginal cost of fishery management is zero. In the context of our projections, this amounts to assuming that the upgrading of water allowed by pollution control will involve insignificant management costs. The "low values" of fishing days involve an assignment of zero

Table 6-3. Changes in Total Values of Fishing Activity per Year with Pollution Control

Scenario	Variants: Sample for days	Method of estimation[a]	Valuation base[b]	Values per day of: Trout fishing	Bass fishing	Rough fishing	Total annual values of: Trout fishing ($10^9)	Bass fishing ($10^9)	Rough fishing ($10^9)	All fishing total ($10^9)
Base	11,000	WLS	Low	$10.96	$ 9.65	$ 7.00	$3.308	$ 7.116	$1.469	$11.893
			High	24.09	21.43	16.03	7.272	15.802	3.363	26.437
	5,000	LOGIT	Low	10.96	9.65	7.00	3.547	7.282	1.346	12.175
			High	24.09	21.43	16.03	7.796	16.171	3.082	27.049
BPT	11,000	WLS	Low	10.96	9.65	7.00	3.500	7.155	1.545	12.200
			High	24.09	21.43	16.03	7.692	15.889	3.539	27.120
	5,000	LOGIT	Low	10.96	9.65	7.00	3.626	7.269	1.385	12.280
			High	24.09	21.43	16.03	7.969	16.143	3.172	27.284
BPT + S	11,000	WLS	Low	10.96	9.65	7.00	3.531	7.162	1.555	12.248
			High	24.09	21.43	16.03	7.761	15.904	3.560	27.225
	5,000	LOGIT	Low	10.96	9.65	7.00	3.643	7.269	1.390	12.302
			High	24.09	21.43	16.03	8.008	16.143	3.182	27.333
BPT + S + pH	11,000	WLS	Low	10.96	9.65	7.00	3.521	7.178	1.557	12.256
			High	24.09	21.43	16.03	7.740	15.941	3.556	27.247
	5,000	LOGIT	Low	10.96	9.65	7.00	3.632	7.288	1.393	12.313
			High	24.09	21.43	16.03	7.983	16.184	3.190	27.357
"BAT"	11,000	WLS	Low	10.96	9.65	7.00	3.552	7.198	1.576	12.326
			High	24.09	21.43	16.03	7.803	15.985	3.610	27.403
	5,000	LOGIT	Low	10.96	9.65	7.00	3.640	7.294	1.403	12.337
			High	24.09	21.43	16.03	8.003	16.199	3.214	27.416

[a]Method for estimation of some-participation probability equations.

[b]Low valuation base = no time cost for travel time; fishery fee treated as equal to constant marginal cost. High valuation base = travel time value at wage rate; fishery fee treated as transfer, not cost. (Effectively assumes there will be insignificant costs of carrying out and sustaining the upgrading of fisheries allowed by pollution control.)

value to travel time and the assumption that the fishery fees entirely reflect resource costs.

Further, in table 6-3 the values per day are multiplied by the days of fishing projected under each scenario. Here we add another element of variation by showing the results for the 11,000 and 5,000-person sample sizes (with WLS and logit, respectively, in estimation) from table 6-1. For both sample sizes the estimated prediction equations for days of fishing per year, by species sought, include the availabilities of all three types of fishable water and the bags of all three species types. For the result of including in the days equations only the water availability and bag numbers for the specific type of fishing involved, see below, table 6-5.

The total annual values estimated range from about $1.3 billion for rough fishing in the base situation (5,000 sample, low valuation of a fishing day) to over $16 billion for bass fishing under the BAT scenario (high valuation of a fishing day). For any combination of assumptions and methods, the total value of rough fishing is about half that of trout fishing while trout fishing's value in turn is about half that of bass fishing. The total value of freshwater recreational fishing (excluding Great Lakes' fishing) is between $12 and $27 billion per year, depending on the assumptions governing the per day values.

The freshwater recreational fishing benefits of the several stages of water pollution control may now at last be derived. This is done in table 6-4 for the four key combinations of methods and assumptions, most importantly those defining the values of fishing days. According to these estimates, the benefits we have been seeking probably lie between $100 and $700 million per year for implementation of BPT. The extra cleanup purchased by going from BPT to BAT is worth somewhere between $60 and $280 million per year.[2] The intermediate additions to BPT (sediment and pH control) together have incremental benefits in the neighborhood of $35 to $130 million;[3] and if those policies were both in place, the addition of BAT would produce incremental fishing benefits of $25 to $160 million per year.

Note that these benefit estimates are all based on the 1975 U.S. population (for the lower 48 states) and on family incomes for that year. In later years (1985 has often been used in previous studies because of its identification with certain objectives of FWPCA(A)/CWA) population and income will both be larger. Thus it is tempting to project higher

Table 6-4. Annual Benefits of Water Pollution Control Accruing via Recreational Fishing

Scenario	Variants: Sample size	Variants: method of estimation[a]	Variants: Valuation base[b]	Total benefits over base ($$10^6$)	Marginal benefits of control increment ($$10^6$)	Total benefits per fisherman[c] ($)	Total benefits per capita (1975 POP)[d] ($)
BPT	11,000	WLS	Low	307	307	5.19	1.45
			High	683	683	11.54	3.22
	5,000	LOGIT	Low	105	105	1.77	0.50
			High	235	235	3.97	1.11
BPT + S	11,000	WLS	Low	355	48	5.99	1.68
			High	788	105	13.31	3.72
	5,000	LOGIT	Low	127	22	2.14	0.60
			High	284	49	4.80	1.34
BPT + S + pH	11,000	WLS	Low	363	8	6.12	1.72
			High	810	22	13.67	3.82
	5,000	LOGIT	Low	138	11	2.33	0.65
			High	308	24	5.20	1.45
"BAT"	11,000	WLS	Low	433	70	7.30	2.05
			High	966	156	16.30	4.56
	5,000	LOGIT	Low	162	24	2.73	0.76
			High	367	59	6.19	1.73

[a]Method of estimation of some-participation probability equations.

[b]Defined in table 6-3, note b.

[c]From table 6-1.

[d]Based on population of lower forty-eight states plus the District of Columbia in 1975 (U.S. Department of Commerce, 1978.

benefits, especially using a participation-based method, by operating on those higher populations. The problem with doing this in the context of the stages of FWPCA(A)/CWA implementation is that the use of technology-based discharge standards means that the higher raw pollution loads almost certainly associated with the larger population and income will be translated into higher discharges and hence poorer ambient water quality. In the context of our study, then, to value the BPT stage of implementation with a large population and income would require going back to our discharge inventory, adjusting raw loads in some complicated way reflecting the distribution of population and the composition of production, reapplying BPT to the adjusted raw loads, repredicting ambient quality and therefore the extent and proportions of fishable water.

Because, however, our scenario 5 is defined by ambient conditions (minimum DO level of 5.0 ppm attained everywhere) the problem of inconsistency does not arise in the same way. If we are willing to: (1) accept that the link between this scenario and BAT must be abandoned; and (2) realize that the costs of attaining the conditions assumed in scenario 5 would increase with the benefits (though not necessarily by the same amount); and (3) ignore the problem that acreage per capita will eventually fall as population rises and the limits of natural waters are reached, we can obtain a 1985 benefit for the scenario 5 conditions based on population change alone. This adjustment requires only a multiplication of our benefit estimate by the ratio between predicted 1985 population for the lower 48 states and the 1975 population figure we used. That ratio, using Census projection series IIC, is 1.09 (U.S. Department of Commerce, 1978). Thus, the adjusted benefits attached to scenario 5 would range from about $205 million to $1.6 million per year.

Returning to the benefit estimates in table 6-4, let us see if the numbers seem plausible. One initial test along these lines is to express the benefits in per capita and per fisherman terms, and this is done in the last two columns of table 6-4 for the total benefits attached to each scenario. The per person (1975 population of the lower 48 states plus the District of Columbia) annual benefits are estimated to be between about $0.50 (for BPT with very conservative valuation assumptions for fishing days) and $4.60 (for BAT with travel time valued at the wage rate and marginal costs of public fisheries assumed to be zero). The corresponding range of values when expressed per angler is from $1.75 to $16.30 per year. These last figures have the mildly encouraging feature[4] of being of the same magnitude as state resident fishing license costs.

Another obvious test for plausibility is comparison of our results with estimates made by others, though such a test must be treated with caution. After all, if our goal of improving methodology is attained, and if our data are reasonably sound, our benefit estimates might be very different from those in the literature for good reason. To pursue this line of comparison, however, let us return to the data presented in chapter 1 and additional data taken from Freeman's interpretive review of pollution control benefit studies (Freeman, 1979). In table 10 of his report (p. 145), Freeman shows freshwater recreational fishing benefit estimates from a study done for the National Commission on Water Quality by the National Planning Association. The numbers that most obviously correspond to some of our results are the NPA/NCWQ benefits for 1985, which assume full implementation of BAT. In 1978 dollars these range from $267 to $677 million per year, and are therefore very close to our results for BAT (scenario 5). This comparison assumes that we can view our results as being expressed in 1978 dollars, a somewhat problematical point as we have noted. Furthermore, Freeman provides his own judgment of the likely range and most likely point estimate for freshwater recreational fishing benefits assuming full implementation of FWPCA(A)/CWA. The range he chooses is from $0.6 to $4.0 billion per year, and his most likely point estimate is $1.0 billion, all in 1978 dollars. In this perspective, our estimates are in and below the low end of the range. Our estimate from the larger sample, based on the assumption of zero marginal cost of public fishery management, and the valuation of travel time at the wage rate is, however, rather strikingly close to Freeman's choice for the most likely point. Whether this should be attributed to the quality of our method, the excellence of Freeman's judgment, or simple luck, is a matter on which readers may be expected to differ. If we were asked, as Freeman was, to supply a best point estimate for the annual benefits of water pollution control accruing via freshwater recreational fishing, we would choose about $400 million for BPT and $600 million for BAT.

The reader will recall, however, that in the course of chapters 4 and 5 we flagged several possible technical variations in our procedure and promised to return to them in chapter 6 and demonstrate at that time the extent to which our choices have influenced the final estimates. We undertake this task in the next section.

VARIATION IN BENEFIT ESTIMATES WITH VARIATION IN METHODS AND ASSUMPTIONS

Table 6-5 shows how the major variations in methods and assumptions described in chapters 4 and 5 affect the benefit estimates. The first and last pairs of rows summarize the results already presented in table 6-4: one set based on the 11,000 sample and the use of weighted least squares in the some-of probability equations; the other based on the 5,000, quasi-proportional sample, with the some-of equations estimated using logit. The row pairs numbered 2 through 6 show variations beginning from the 11,000 sample and the use of weighted least squares. In the second pair of rows the effect of outlier deletion on the days-equations estimation is shown to be a significant reduction (by roughly one-third) in the benefits for every scenario and for both high and low valuation assumptions. For the third row pair, we use the 5,000 sample in the days equation estimation (with outliers in the data set) and produce a set of estimates that are uniformly the highest across all valuation and scenario pairs. The next two rows (4 and 5) show how restricting the availability and bag measures in the days equations to the pair particular to the type of fishing in question (for example, only trout share and trout bag in the trout days equations) affects the estimates in rows 1 and 3 respectively. That effect is, uniformly, a substantial reduction across scenarios and for both valuation assumptions. Row pair 6 involves further reduction in the sample size used in the days equation (to 1,000) and should be compared with row pairs 1 and 3. The smallest sample produces benefit estimates lower than either the 11,000 or 5,000 samples. Finally, for row pair 7, we depend entirely on the 5,000 quasi-proportional sample, but use OLS in estimating the some-of participation probability equations. The resulting estimates are modestly larger than those found in the last row pair, which gives us the uniformly lowest benefit estimates for all combinations of scenario and valuation assumptions.

The general lesson of table 6-5 is that sample design, methodology, variable specification and valuation assumptions all make substantial differences in the benefit estimates finally produced. For any single choice of valuation assumptions, high or low, and any particular scenario, the ratio of highest to lowest estimate in the table is about three. When we look across valuation assumptions as well, that ratio rises to ten! Thus, one must conclude that substantial uncertainty remains in

Table 6-5. Variations in Benefit Estimates with Sample, Methodology, and Assumptions

Case and valuation base[a]	Source sample, estimation method, treatment of explanatory variables and outliers in days equations: Probability of "some" fishing equations	Intensity of fishing equations	Scenario and valuation base (total benefits in $\$10^6$): Scenario I LO	Scenario I HI	Scenario II LO	Scenario II HI	Scenario III LO	Scenario III HI	Scenario IV LO	Scenario IV HI
1. LO	11,000 WLS[b]	11,000 WLS, All, All[d]	307		355		363		433	
1. HI				683		788		810		966
2. LO	11,000 WLS	11,000 WLS, All, Outlier Delete[e]	216		251		258		308	
2. HI				479		556		572		682
3. LO	11,000 WLS	5,000 OLS, All, All[g]	362		417		426		507	
3. HI				804		926		946		1128
4. LO	11,000 WLS	11,000 WLS, 1 AC/1 BAG, All[f]	238		279		291		348	
4. HI				529		618		645		772
5. LO	11,000 WLS	5,000 OLS, 1 AC/1 BAG, All	260		304		315		376	
5. HI				577		674		700		836
6. LO	11,000 WLS	1,000 OLS, All, All[h]	211		249		262		314	
6. HI				465		549		578		693
7. LO	5,000 OLS[c]	5,000 OLS, All, All	122		146		156		186	
7. HI				272		326		349		416
8. LO	5,000 LOGIT	5,000 OLS, All,All	105		127		138		162	
8. HI				235		284		308		367

[a]For definition of low and high valuation bases see table 6-3, ncte b.

[b]The 11,000 sample refers to the full sample of usable fishermen mail survey responses from NSHFWR. "WLS" means weighted least squares.

[c]The 5,000 sample refers to a quasi proportional sample constructed using a piecewise linear program as described in appendix B.

[d]11,000 sample as in note b. The first "all" refers to the number of acreage and bag variables in the day's equation: in the "all" cases, all acreage and bag for all three fishing types appear in each of the three equations. The second "all" refers to the use of all available observations.

[e]"Outlier Delete" means that the largest 5 percent of the reported days' observations have been deleted for each fishing type.

[f]"1 AC/1 BAG" means that only the acreage and bag variables specific to the type of fishing day in question appear in the days equations.

[g]"5,000" refers to the quasi-proportional sample as in note c.

[h]"1,000" refers to the perfectly proportional sampie.

the estimates of freshwater recreational fishing benefits, though as we have observed, not all of the combinations in the table have an equal claim to our attention.

Recall that in addition to the sources of uncertainty represented in table 6-5 there are at least two other potentially serious problems. First, as pointed out in chapter 2, the central source of information on quality (availability) changes, the RFF Water Quality Network Model, in its current stage of development, offers significantly less than complete coverage of U.S. freshwater acreage. Such coverage as it does provide is necessarily approximate quantitatively and narrow in the measures offered. Second, the discussions of chapter 1 and appendix A demonstrate that the use of pre-policy average consumer's surplus to value the policy-related change in participation gives only an approximation to the quantity we are seeking but cannot observe. This approximation may be a very good one or may be off by a factor of two or three in either direction and we cannot be sure which situation obtains because of our inability to measure the elasticity of the underlying demand function(s) for fishing days. Further, we cannot even assume that the approximation is equally good for cold-water game fish, warm-water game fish, and rough fish. Because it appears, as discussed in the next section, that the angler responses to availability are different across fishing types, our method might come close to the correct value for one fishing type and be off by a factor of two for another.

We return below to the lessons we think it appropriate to draw from these sobering observations. At this point, however, one final technical question requires clarification.

ANALYSIS OF THE SOURCES OF BENEFITS AND A POSSIBLE ALTERNATIVE ESTIMATION ROUTE

An examination of the intermediate results summarized in tables 6-1 and 6-3 may suggest that the estimation of recreational fishing benefits should not have to involve some of the complications we have stressed. In particular, the benefits we find seem to come in very large part from a simple increase in the days participation predicted for fishermen who were fisherman before any quality change ("old fishermen" or OF). The increase in the number of fishermen and the differential increase in game fishing participation do not seem to play any significant role. Thus, let:

OF = the total number of fishermen before pollution control

NF = the additional fishermen attracted by new available water

OD = the total days of participation per average fisherman per year, undifferentiated by species sought, before pollution control[5]

ΔD = the increment to OD predicted from pollution control

OV = the weighted average value of a fishing day, where the weights are the shares of prepollution control fishing days accounted for each of the day types

Using this notation the total benefits of any state of control may be approximated as:

$$B = \left[OF \times \Delta D + NF \times OD + NF \times \Delta D\right] OV$$

Consider, for example, the benefits estimated for scenario 5, using the 11,000 person sample in the fishery-type participation and days-equations steps, and the high valuation base. Then:

$$OF = 59.16 \times 10^{6}$$

$$NF = 0.12 \times 10^{6}$$

$$OD = 21.11$$

$$\Delta D = 0.74$$

$$OV = \$21.17$$

$$B = \left[59.16 \times 0.74 + 0.12 \times 21.11 + 0.12 \times 0.74\right](21.17)(10^{6})$$

$$= \left[43.78 + 2.53 + 0.09\right](21.17)(10^{6})$$

$$B = \$982 \times 10^{6}$$

Thus, 94 percent of the benefits are attributable to the increased days fished by the "old fishermen." None of the benefit estimated in this way is attributable to the upgrading of species, for the valuation used is the base-weighted average.

Why not then do away with the elaborate superstructure of fishery classes and fisherman participation in each class? Why not go directly to the heart of things and estimate total fishing days in two steps: (1) the fish/do-not-fish decision, (2) the decisions about undifferentiated days of fishing per year? Both these relations could be estimated using total fishable acres per capita as the key independent variable.

Why not, indeed. We explored this possibility after making the above analysis, and found that undifferentiated total days per year per fisherman are negatively related to the total availability per capita of fishable water. To see how this can be so and why this apparent short cut tangles us hopelessly in the aggregation problem, consider a simple, artificial example involving nine "states," in each of which only one kind of fishery and only that kind of fishing activity (days) is observed. Label the states 1,...,9 and let the observations be as summarized in table 6-6. The relationships in figure 6-1 between acres per capita and days are positive for each fishery class.

Table 6-6. A Simple Example: Fishery Types and States

	TROUT		BASS		ROUGH	
State	Days	Acres/cap	Days	Acres/cap	Days	Acres/cap
1	20	1				
2	10	0.5				
3	30	1.5				
4			10	2		
5			5	1		
6			15	3		
7					5	3
8					2.5	1.5
9					7.5	4.5

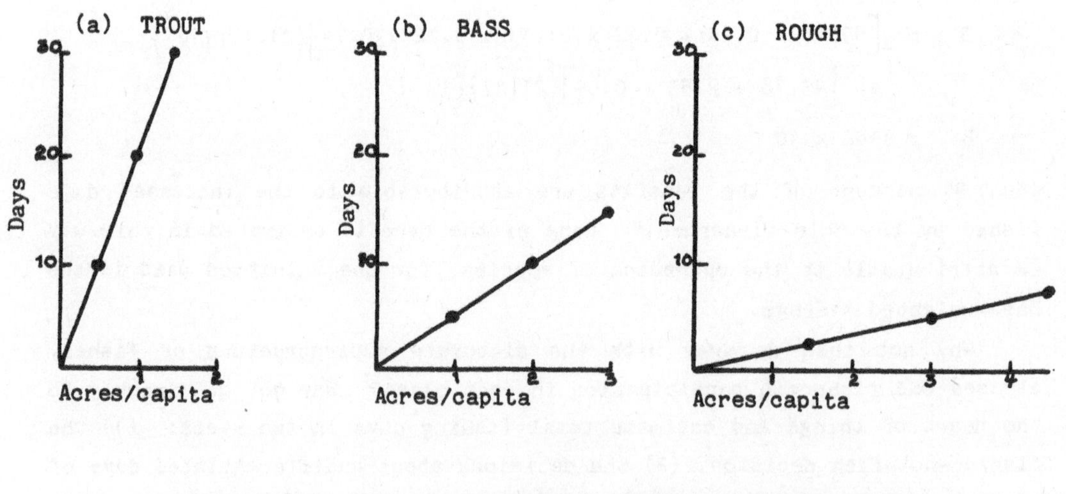

Figure 6-1. The simple example plotted by fishery type

When the data are taken together, without any attempt to distinguish fishery or type of fishing day, however, the resulting plot in figure 6-2 looks rather odd.

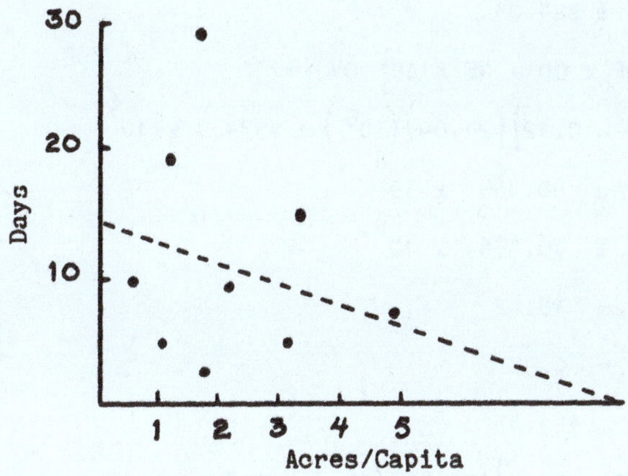

Figure 6-2. The simple example: An aggregate plot

When one fits a linear relation to these points using OLS, the equation is:

$$\widehat{DAYS} = 14.37 - 1.35(AC/CAP)$$

and is shown in the figure by a dashed line. A negative slope has resulted from aggregating several positive relations.

Thus, it is not possible to simplify the benefit estimation procedure for recreational fishing by eliminating the species class superstructure. Even though the upgrading phenomenon turns out to be much less important than we anticipated, the fishery classifications cannot be ignored because the way in which participation decisions are made in each class differs. Aggregation of days and acres therefore amounts to mixing observations from different "worlds," and the resulting relations are meaningless.

To reinforce this conclusion, it will be worthwhile to look at the results of repeating the benefit analysis, performed above at the aggregate level, for each of the fishery types in turn. For the same combination of methods and assumptions the values and calculations are as follows:

Trout:

$$OF = 21.937 \times 10^6$$
$$NF = 1.728 \times 10^6$$
$$OD = 13.76$$
$$\Delta D = -0.06$$
$$OV = \$24.09$$

$$B = \left[OF \times \Delta D + NF \times OD + NF \times \Delta D\right] OV =$$
$$\left[-1.54 + 23.85 - 0.12\right](24.09)(10^6) = \$534.6 \times 10^6$$

Bass:

$$OF = 40.465 \times 10^6$$
$$NF = -0.356 \times 10^6$$
$$OD = 18.22$$
$$\Delta D = 0.38$$
$$OV = \$21.43$$

$$B = \left[15.38 - 6.65 - 0.14\right](21.43)(10^6) = \$184.1 \times 10^6$$

Rough:

$$OF = 20.70 \times 10^6$$
$$NF = 0.39 \times 10^6$$
$$OD = 10.14$$
$$\Delta D = 0.54$$
$$OV = \$16.03$$

$$B = \left[11.18 + 3.91 + 0.21\right](16.03)(10^6) = \$254.3 \times 10^6$$

$$\text{TOTAL } B \cong \$964.0 \times 10^6$$

Notice that the pattern is different for each fishery. For trout, the benefits result entirely (more than entirely because of the negative terms) from the increase in anglers. Here is the upgrading effect in action. For bass and rough fishing, it is the increase in days per angler that is important. But for bass, this effect more than accounts for the calculated benefits, there being a fall in the number of bass fishermen; while for rough fish, there is also a positive contribution from the increasing number of anglers. Thus, a more careful analysis shows us in another way that the benefit estimates do not, in any real sense, derive only from increases in days fished. The pattern varies across fisheries, and only trout fishing displays the pattern of upgrading participation that we had expected to see more generally.

CONCLUSIONS

It seems that we may have here the research analogue of the half-full glass of water. The data we had available about policy effects on the natural world, and about individual choices in the face of varying availability of recreational fishing resources, were outstanding judged against those used in other studies of water pollution control benefits. We have tried consistently to use sound methodology and have explored the implications of alternative choices where soundness was not a black and white matter. At the end, however, we find that our benefit estimates for a particular level of policy implementation can vary by an order of magnitude from lowest to highest. Is the glass half full--a record of methodological improvement? Or half empty--a lesson that these methods are inevitably doomed to fail in the eyes of those who must make and justify policy? Let us consider in turn the positive and negative assessments.

A Positive View

Where interest is in the benefits of environmental programs or regulations that apply to substantial geographic areas and involve comprehensive change in the availability of amenity resources, an estimation method based on individuals and their participation choices has substantial advantages. The most important of these is that it bypasses the problems inherent in trying to aggregate site-related benefits. These problems, in turn, arise because the universe of "sites" must inevitably remain ill-defined, there being no entirely satisfactory way of identifying the set of mutually exclusive sites that together "exhaust" the available resource. In addition, to the extent that a satisfactory set of sites can be defined for a problem, the changes occurring in general at all sites, though in different amounts at each, create a difficult general equilibrium problem to be solved before consumer response and hence benefits can be predicted.

For the participation method to be workable, it is necessary to have the following key pieces of information:

> a measure of resource availability that both makes sense in recreation-related terms and that can be connected to the policy program, or regulation of interest

a substantial amount of data on the participation decisions of individuals, data that allow us to connect the pre-policy decisions to the pre-policy levels of resource availability

a basis for valuing whatever change in participation is forecast on the basis of the policy-related change in resource availability

For any attempt to estimate the benefits of an environmental quality improvement, we must have some way to translate the policy into suitably comprehensive measures of the resulting changes in ambient environmental quality. This is not unique to a participation method. What is crucial for participation is that we can measure and predict ambient quality in terms related to participation decisions. In this study, for example, the key concepts were fishability and dominant fish species supportable.

While the method does depend on the existence of consumer (household production) decisions based on resource availability, those decisions do not have to concern water contact, or even active sports. If, for example, we could show that more hiking or more picnicking would be done if water quality improved, the method would allow us to obtain an estimate of the benefits accruing in those ways.[6] Nonetheless, the method probably has its greatest payoff where active participation in sports directly connected to water are part of the benefit-generating system. The examples of possible extensions suggested below are offered with that in mind.

One such extension would be to the estimation of the benefits of water pollution control accruing via recreational fishing in estuarine and coastal waters. Most of the necessary pieces of information are already available, though one very central one is not. In particular, the same surveys from which we drew our participation data include questions on saltwater fishing. Analysis of the willingness to pay questions from those surveys has allowed others to produce value estimates for days of saltwater fishing of various specific types (Charbonneau and Hay, 1978). What is missing is the link between discharges and ambient water quality measured in terms meaningful to fishermen. Though a comprehensive inventory of coastal zone pollution sources is being constructed at Resources for the Future, it is not now linked to ambient water quality models. Nor is it clear what measures of quality would correspond to the freshwater fishable-nonfishable and dominant species classifications used in this study. It seems unlikely that any open ocean waters are actually unfishable, nor are we aware of classifications corresponding to the rough fish/game fish split along water quality lines.

A broader field for application of this method is that encompassing other kinds of water-based recreation. Here there is much less by way of available data, and a really satisfactory effort would require survey work on participation as well as development of value measures and of the appropriate quality links. (However, see Harrington, 1981, for an effort to make use of available survey data.) The survey data base would necessarily be expensive because of the large samples needed for robust estimation. In addition, the survey instrument would have to be fairly complicated if an attempt were to be made to cover boating, swimming, wildfowl hunting, and such uses as hiking and picnicking, which have water as a backdrop. The lesson of the aggregation example explored earlier in this chapter is that it would be necessary to carefully separate out uses that, while superficially the same, might be expected to involve different responses to water quality and quantity conditions. For example, "boating" would probably not work out any better than "fishing" as a category. We would expect to have to separate out canoeing, rafting, power boating, sailboating and perhaps other kinds as well. A quality indicator would be necessary for each.

Another possible application of this method is to the determination of the damages accruing to the nation from acidification of lakes, ponds, and streams via lost recreational fishing. While this acidification is widely thought to be the result of man's discharges of sulfur and nitrogen compounds to the atmosphere, the linkage from discharges to acid precipitation and dry deposition remains a matter of considerable dispute. Though it would not, therefore, be possible to estimate the benefits of any particular program aimed at changing discharge, there is still room for a sound measure of the recreational fishing damages as part of the effort to assess the size of the problem.

The missing link for this application is a comprehensive inventory of water bodies suffering acidification from what are believed to be nonnatural (and non-mine-drainage) causes. No such inventory exists at present. Additionally, information on the past or potential future fishery classification of each water body would be necessary. With these data to supplement those already available on fishable water by extent and classification, on participation in fishing activity, and on the valuation, of fishing days, it would be possible to develop an estimate of the increase in value of national freshwater recreational fishing that would be predicted to result from a change in fishery status (or in predicted daily bags) for currently acidified waters to their natural fishery class.

Overall, then, we believe that the method developed and illustrated in this study is potentially very useful when national benefit estimates are wanted for categories of environmental services involving participation choices by individuals.

Some Negative Considerations

There are two problem levels that require candid treatment in order to put in proper perspective the positive commments and suggestions just made. First, we must briefly review the shortcomings of this study, not simply as an act of contrition, but becuse some of these shortcomings will help us to identify the second, deeper, problem level: those difficulties that are inherent in the method and the recognition of which must lead us to look beyond these techniques in the search for truly comprehensive benefit estimates.

While we have been as careful as time and budget have permitted, there remain significant problems with data and models, which imply that it would be foolish to claim that our results represent the last word.

One fundamental difficulty is the roughness of the connection between pollution discharges and predicted ambient conditions expressed in terms of fishery class. This roughness has several facets:

> The coverage of U.S. waters in our water quality network model is far from complete overall and is differentially spotty across the states.
>
> Only the BOD-DO part of the discharge-ambient quality linkage is covered in the model.
>
> The definition of discharge points and reaches within which perfect mixing is assumed to occur creates another layer of approximation in the model.
>
> The mapping from ambient quality in physical/chemical terms to species classes introduces further approximation because of the broadness of our classes and the imperfections of our knowledge of the factors influencing species survival and successful competition.

Because this linkage seemed to perform better when aggregation to the national level washed out the state-by-state variations, we used national acreage aggregates in our benefit estimation work. This may be the best course in the circumstances, but one would prefer to build national estimates up from the states when nonlinear equations are involved.

Our data on participation did not allow us to identify the place of residence of the individual respondents, nor did the questions ask about the specific places at which individuals fished. Coupling these gaps with our incomplete model coverage of state waters, we find ourselves completely unable to take account of any direct measure of accessibility and hence of changing accessibility due to pollution control. Quantity stands as an imperfect proxy for accessibility. It is impossible to say a priori how, or even whether, improvement in this respect would affect the estimates.

Our estimates of average willingness to pay for fishing days in the several classes, while broadly consistent with the results obtained by others, are based on data from a sample of fishing sites which turned out to have many members with characteristics quite different from those of the free, public, and largely natural sites we usually have in mind when we speak of recreational fishing. The surveyed sites appeared to stress bag and to provide an ambience in which the hand of man was not hidden. We suspect that the result of this difference is to bias down the value estimates we derive and hence to pull down the overall benefit estimates as well.

Further, the travel cost method of obtaining a value for average consumer's surplus suffers from its own faults. These include extreme sensitivity to the choice of functional form in the initial step of relating visits (or visits per capita) to travel cost; and sensitivity as well to decisions made about the treatment of visits (visitors) from very distant zones and of distant zones from which no visits are recorded.

Finally, we have not been able to take account of the benefits that may accrue via the increase in enjoyment (the higher valuation) of every fishing day due to cleaner water.

The above problems, while specific to this study, are very likely ubiquitous. The indirect methods of benefit estimation, with all their accumulated theoretical baggage so beloved by economists, are enormously data hungry. This is true even when we move away from the theoretically correct base (the household production function approach with marginal cost and marginal willingness to pay) and step, as we must, into the world of reduced forms, proxies for availabililty, and separation of valuation from prediction of quantity changes (Bockstael and McConnell, 1981). These compromises and approximations still leave us with the need for very large and very carefully crafted surveys in order that participation decisions and valuation may be inferred from data that suffer as little as possible from the taint of strategic, self-interested behavior.

Is it wisdom or professional myopia that encourages the call for yet more and more expensive data to be mercilessly subjected to the latest and most sophisticated econometric weapons? After having participated in this sport ourselves, we are inclined to think that an examination of alternative routes is needed. Economists by and large take very seriously the a priori arguments that, where public goods are concerned, direct demand revelation can only be managed through very complex procedures designed to neutralize strategic biases (e.g., Green and Laffont, 1972.) Whether the difficulties of eliciting useful responses to direct questions about willingness to pay for environmental quality programs are any greater than those involved in teasing answers to theoretically sound but quite artificial questions out of retrospective surveys of individual leisure-time activities is at least an open question. In short, we believe that the payoffs to experimentation with direct (survey) methods are likely to be greater than those to further elaboration of the already quite tortured reasoning and techniques behind the respectable indirect methods. Certainly if the costs of benefit estimation are to be considered, and if the current emphasis on attributing program benefits to the component regulations is to be continued, the direct methods deserve attention.[7] We may never be able to produce believable comprehensive program benefits via indirect methods, and the attempt to do so will be very, very expensive, for it will require new data bases, not just new regression runs. If millions of dollars of survey money are to be spent, it might be more efficient to spend part of it on substantial, national, willingness to pay surveys.

NOTES

1. There is also a problem of allocating the survey responses for days of warm-water by type of fish sought to our "bass" and "rough" categories. In the NSHFWR survey, there were two questions asking about warm-water fishing days spent in 1975. The second asked for days differentiated by species, but, if more than one species was sought on any day, the respondent was not to apportion the day across species categories. Instead, each species category was to be treated as if a separate day was spent. This produced double counting. In order to avoid double counting of days but still apportion warm-water days across species categories, we calculated the fraction of total double-counted warm-water days spent seeking each species and applied these fractions to the total days (not double counted) from the first question. This may have produced systematic errors in the "days" variables.

2. We remind the reader that it is difficult to pin down a year to attach to these dollars. The travel cost per mile used in chapter 5 was given by the American Automobile Association for 1980 but it is lower than that used by several other researchers in studies done in the mid 1970s. Income data for the travel cost zones is from 1975 and visitation questions were asked about the 1979 season. We are inclined to think that 1978 is a reasonable choice when, as below, we need a specific base.

3. The incremental benefits attributable to pH control appear very small, but two cautions are in order. First, this applies only to correction of acid mine drainage problems, not to reversal of acidification usually attributed to acid precipitation and dry deposition. Second, we suspect that our inventory of streams and lakes suffering from acid mine drainage was a good deal less than complete. Hence, even that effect is probably underestimated here.

4. This is only mildly encouraging because there is no argument to lead one to suspect such a result; and data on the erratic relationship between licensed fishermen by state and numbers of people who say they fish suggest that we should not make too much of anything having to do with state licenses.

5. This is a fisherman-weighted average; that is, total days spent divided by total fishermen.

6. Notice that how the participation information is gathered makes a difference. It must be gathered via comprehensive, off-site interviews in which at least some of the questions ask simply about participation, untied to specific places or to specific water quality levels at specific places. On-site interviews will inevitably tangle the data in the general equilibrium problem, as will questions that attempt to tie decisions to specific quality levels at specific places.

7. For one useful review of the state of the art in willingness to pay survey methods (sometimes called hypothetical or contingent valuation), see Brookshire and Crocker (1981). For discussion of possible biases, their interrelation, and practical seriousness in the context of a water pollution benefit project, see Mitchell and Carson (1981).

REFERENCES

Bockstael, Nancy E. and Kenneth McConnell. 1981. "Theory and Estimation of the Household Production Function for Wildlife Recreation," Journal of Environmental Economics and Mangement vol. 8, no. 3 (September) pp. 199-214.

Brookshire, David S., and Thomas D. Crocker. 1981. "The Advantages of Contingent Valuation Methods for Benefit-Cost Analysis," Public Choice vol. 36, pp. 235-252.

Charbonneau, John and Michael J. Hay. 1978. "Determinants and Economic Values of Hunting and Fishing." A paper presented at the 43rd North American Wildlife and Natural Resources Conference, Phoenix, Arizona, March 18-22, 20 pp.

Freeman, A. Myrick III. 1979. The Benefits of Air and Water Pollution Control: A Review and Synthesis of Recent Estimates. A report prepared for the Council on Environmental Quality, Washington, D.C.

Green, J., and J. J. Laffont. 1972. "Characterization of Satisfactory Mechanisms for the Revelation of Preferences for Public Goods," Econometrica vol. 45 (March) pp. 427-438.

Harrington, Winston. 1981. "The Distribution of Recreational Benefits from Improved Water Quality: A Microsimulation." Discussion Paper No. 80 (Washington, D.C., Resources for the Future).

Mitchell, Robert C., and Richard Carson. 1981. "An Experiment in Determining Willingness to Pay for National Water Quality," Unpublished draft report to EPA (Washington, D.C., Resources for the Future.)

U.S. Department of Commerce, Bureau of the Census. 1978. Statistical Abstract of the United States: 1978, (Washington, D.C., U.S. Government Printing Office) p. 16.

Appendix A

VALUATION IN THE CONTEXT OF THE HOUSEHOLD PRODUCTION FUNCTION MODEL

In household production function theory, the consumer produces final service flows--recreation days--using consumption technology, time, purchased goods inputs, and nonmarket goods inputs. The individual's self-supply equation for a given constant returns-to-scale technology defines the marginal cost of producing a service flow q (Deyak and Smith, 1978):

$$mc = \partial c/\partial q = f\ (p,a,w) \tag{1}$$

where

mc = individual's marginal cost
c = total cost
q = recreation days, the service flow
p = prices of market goods
a = resource availability
w = wage rate

Each individual's inverse demand function can be expressed in terms of income and tastes. If WP is the individual's marginal willingness to pay for the service flows, then we write:

$$WP = \frac{\partial TWP}{\partial q} = g\ (y,t,q) \tag{2}$$

where

y = individual's income
t = individual's tastes
TWP = total willingness to pay

Individual equilibrium is determined where mc = WP. Two reduced-form equations can be produced from the structural equations in (1) and (2); one for quantity and one for price:

$$q = \Upsilon(y,t,p,a,w) \tag{3}$$

$$mc = WP = \phi(y,t,p,a,w) \tag{4}$$

Reduced-form equation (3) is the one we estimated (in parts) in chapters 3 and 4. Structural equations (1) and (2) were not estimated, primarily because it is usually impossible to produce observed values of mc.[1]

The conceptually correct measure of benefits arising from increased availability may be written in terms of the above structural equations. Consider any individual, i, and recognize that his marginal cost of producing the recreation experience is a function of availability:

$mc^i(a_o)$ = individual i's marginal cost at pre-policy availability a_o particular to i.

$mc^i(a_1)$ = individual i's marginal cost at post-policy availability a_1 particular to i.

For this individual the net benefit of the policy, $nb^i(a_o, a_1)$ can be written as:

$$nb^i(a_0, a_1) = \int_{q(mc^i(a_0))}^{q(mc^i(a_i))} WP(q)\,dq + q(mc^i(a_0))\left[\left(mc^i(a_0) - mc^i(a_1)\right)\right] - mc^i(a_1)\left[q(mc^i(a_1)) - q(mc^i(a_0))\right]$$

This expression is depicted graphically for two individuals in figure A-1.

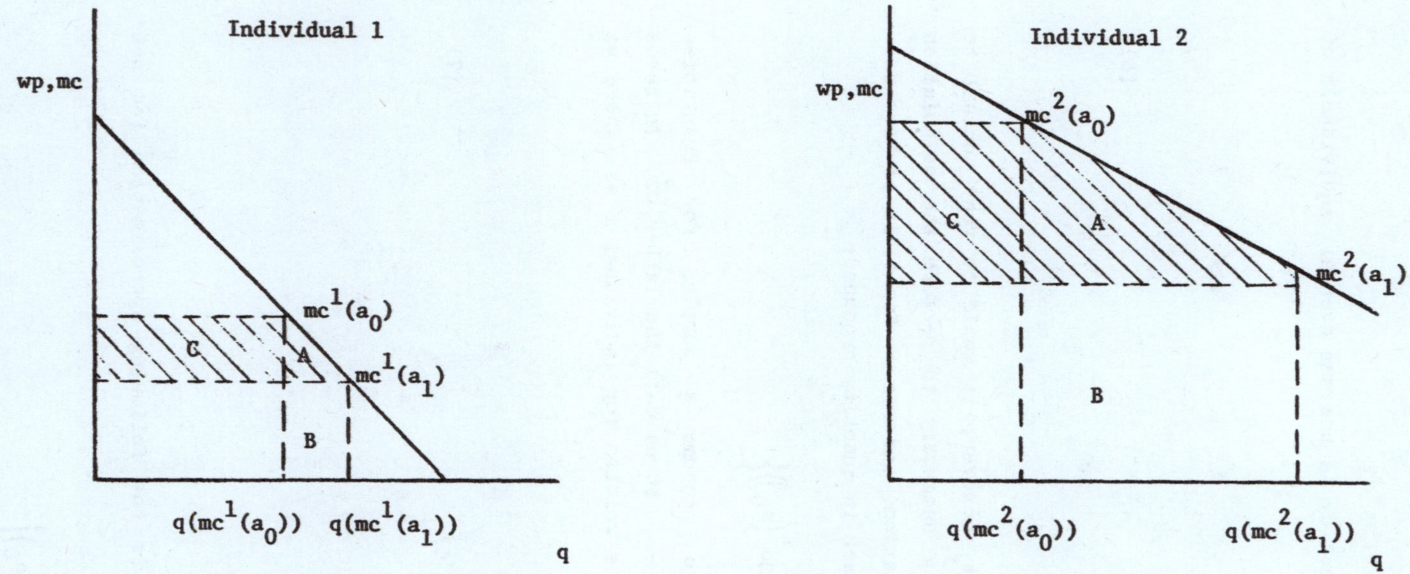

<u>Legend</u>

- $nb^i(a_0,a_1) = C+A = A+B+C-B =$ [hatched area]
- $\int_{q(mc^i(a_0))}^{q(mc^i(a_1))} wp(q)dq = A+B$
- $(q(mc^i(a_0)))(mc^i(a_0)-mc^i(a_1)) = C$
- $\left(mc^i(a_1)\right)\left(q(mc^i(a_1))-q(mc^i(a_0))\right) = B$

Figure A-1. Individual net benefits of increased availability.

The aggregate net benefit of the policy is the sum over all individuals of the net benefits in (5):

$$NB = \sum_{i=1}^{N} nb^i\left(a_0,a_1\right) \tag{6}$$

As we have noted in the text, however, it would be very unusual to have the detailed individual data necessary to perform the calculation written so easily. A common situation, and one we faced, is to have data allowing a prediction of the increase in quantity produced:

$$\sum_i \left\{ q^i\left(mc^i\left(a_1\right)\right) - q^i\left(mc^i\left(a_0\right)\right)\right\}$$

and a measure of the average consumer's surplus for quantities produced/consumed before the change. Let us call the latter $\overline{CS}$. In terms of our demand expression it may be written for individual i as (where we simplify the expressions for q^i):

$$\overline{CS}^i = \int_0^{q^i\left(a_0\right)} W^P\ (q)dq/q\left(a_0\right) \tag{7}$$

Then

$$\overline{CS} = \sum_i \overline{CS}^i/N$$

Under what circumstances, then, is the following approximation for net benefits a good one?

$$\widetilde{NB} = \overline{CS} \sum_i \left[q^i\left(a_1\right) - q^i\left(a_0\right)\right]$$

We have examined this question, using a representative consumer's situation, for the simplest demand function, a linear one, $p = a + bQ$, and for a constant elasticity function with elasticity η, $p = A^{1/\eta} Q^{-1/\eta}$. If the ratio $q(a_1)/q(a_0)$ is written as k, the following expressions may be obtained:

linear demand: $$\frac{\widetilde{NB}}{NB} = \frac{1}{1+k} < \frac{1}{2}$$

constant elasticity demand: $$\frac{\widetilde{NB}}{NB} = \left(1 - \eta q_0^{\frac{1}{\eta} - 1}\right)\left(\frac{1-k}{1-k^{1-1/\eta}}\right)$$

where $q_0 \equiv q(a_0)$

The results of calculation using this latter formula are summarized in figure A-2 for $q_0 = 10$, where the ratio $\widetilde{NB}/NB$ is on the vertical axis and $k = (q_1/q_0)$ is on the horizontal axis. Note that the goodness of the approximation is crucially dependent on η, the elasticity, and quite insensitive to k.[2] It appears that the actual factor, k, for U.S. pollution control policies is probably close to one. And one can argue that for any specific type of fishing, price elasticity should be rather high, because of the availability of substitutes. These observations suggest that the approximation may be a good one. On the other hand, the arc elasticities of the site-demand functions derived in chapter 5 are in the neighborhood of 0.6 to 0.7. If this evidence can be taken as even roughly indicative of the elasticities of activity demand functions, the benefits will be quite seriously overestimated by the approximation. Clearly, there is a need for sound estimates of activity demand functions.

DERIVING THE FORMULAE

For a linear demand function, $p_0 = a + bQ$, the true measure of NB, for a change in availability captured by a downward shift in a horizontal marginal cost curve from $p_0 \equiv WP(a_0)$ to $p_1 \equiv WP(a_1)$ is given by:

$$NB = (p_0 - p_1)\left(\frac{q_0 + q_1}{2}\right) = \frac{b(q_0 - q_1)(q_0 + q_1)}{2}$$

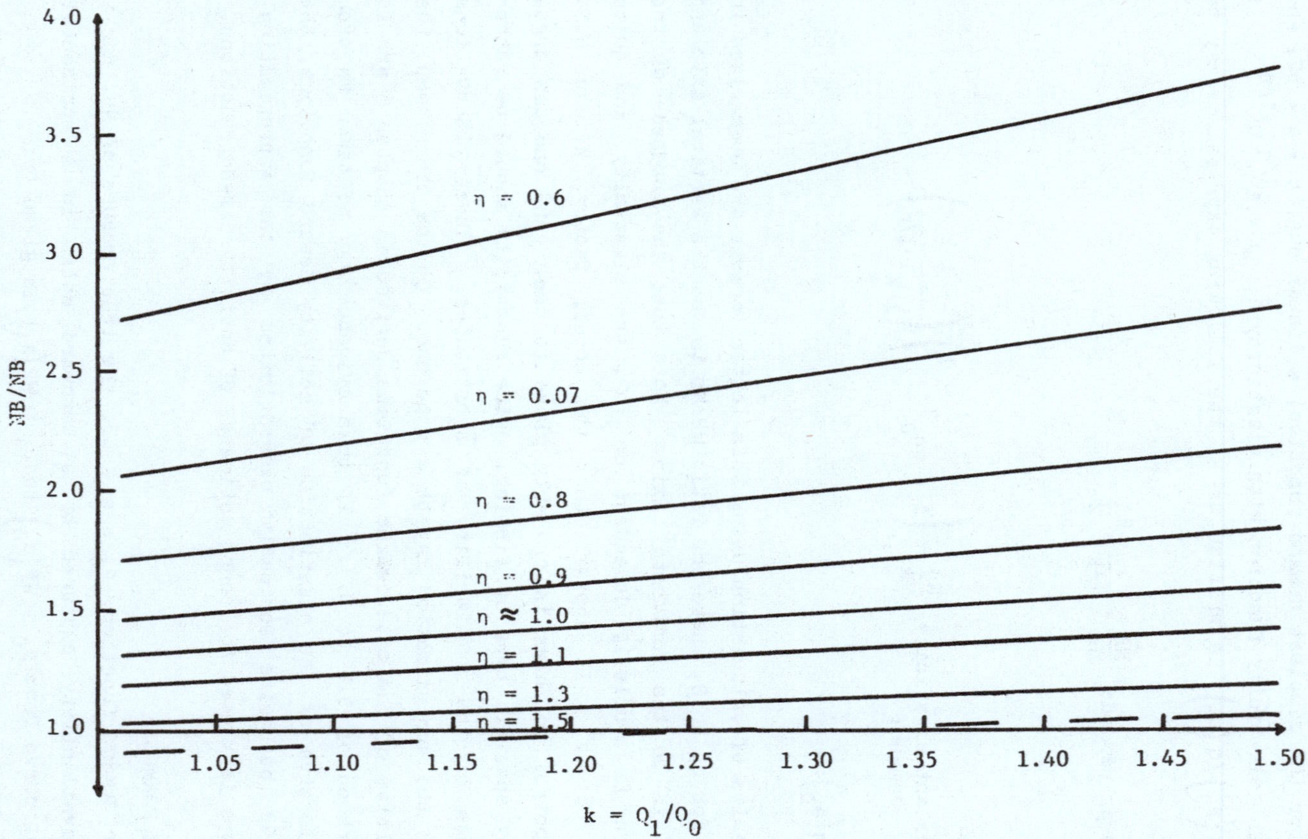

Figure A-2. $\widetilde{NB}/NB$ for the representative recreationist for a variety of demand elasticities (η), as a function of the ratio of post-to-pre-policy days spent per year, given $Q_0 = 10$.

The approximation $\widetilde{NB}$ is given by the product of average consumer's surplus in the base situation:

$$\overline{CS} = \frac{(a-p_0)(q_o)}{2q_0} = \frac{a-p_0}{2}$$

and the change in quantity demanded, $q_1 - q_0$. The ratio $\widetilde{NB}/NB$, is, then

$$\frac{\widetilde{NB}}{NB} = \frac{bq_0(q_0 - q_1)/2}{b(q_0 - q_1)(q_0 + q_1)/2} = \frac{q_0}{q_0 + q_1} = \frac{1}{1 + k}$$

The derivation for the constant elasticity function, $p = A^{1/\eta}Q^{-1/\eta}$, is messier and involves an approximating truncation of the demand curve in order that the integral remain bounded in the derivation of average consumer's surplus in the base case. The true net benefit produced by the vertical shift in availability given by $p_0 \longrightarrow p_1$, is:

$$NB = \int_{p_1}^{p_0} Ap^{-\eta}dp = \frac{A}{1-\eta}\left[p_0{}^{-\eta+1} - p_1{}^{-\eta+1}\right] = \frac{A^{1/\eta}}{1-\eta}\left[q_0{}^{1-1/\eta} - q_1{}^{1-1/\eta}\right]$$

Average consumer's surplus in the base case in approximated by:

$$\overline{CS} = \left[\int_1^{q_0} \left(A^{1/\eta}q^{-1/\eta}\right) dq - p_0q_0\right]\frac{1}{q_0}$$

$$= A^{1/\eta}\left[\left[\frac{q^{1-1/\eta}}{1-1/\eta}\right]_1^{q_0} \left(q_0{}^{-1}\right) - q_0{}^{-1/\eta}\right]$$

$$= \frac{A^{1/\eta}}{\eta-1}\left[q_0{}^{-1/\eta} - \eta/q_0\right]$$

And since $\widetilde{NB} = \overline{CS}\left(q_1 - q_0\right)$, we have:

$$\frac{\widetilde{NB}}{NB} = \frac{\frac{A^{1/\eta}}{\eta-1}\left[q_0^{-1/\eta} - \eta/q_0\right]\left[q_1 - q_0\right]}{\frac{A^{1/\eta}}{1-\eta}\left[q_0^{1 - 1/\eta} - q_1^{1 - 1/\eta}\right]}$$

Substituting kq_0 for q_1, and simplifying, we obtain

$$\frac{\widetilde{NB}}{NB} = (-1)\left[\frac{\left[q_0^{-1/\eta + 1} - \eta\right]\left[k - 1\right]}{q_0^{-1/\eta + 1}\left[1 - k^{1 - 1/\eta}\right]}\right]$$

$$= \left[(1 - \eta q_0^{1/\eta - 1}\right]\left[\frac{1 - k}{1 - k^{1 - 1/\eta}}\right]$$

NOTES

1. Deyak and Smith (1978) suggest that using a proxy variable such as distance traveled for mc is unsatisfactory.

2. Note that for the unit elasticity case the ratio $\widetilde{NB}/NB$ is not defined.

Appendix B

DESIGNING "ALMOST-PROPORTIONAL" SAMPLES FOR LOGIT ESTIMATION

As noted in the text, the NSHFWR mail survey contains serious overrepresentation of small states and of rural areas in all states. This distorts the participation equation estimates. To avoid the problems raised by this, we have two choices: weighted regression, which we use with least-squares, linear equations; and the construction of one or more subsamples reflecting proper or nearly proper weights. The largest properly proportional subsample contains only 1,172 observations, but there may exist one or more subsamples of size $1,172 < N < 11,168$ which can be shown to be "not too bad" as representations of the population distribution of fishermen. One rule by which we could recognize "not too bad" would be to minimize the sum of the squared deviations between the actual and the desired cell frequencies. This rule defines a quadratic programming problem, where the constraints arise from the availability of observations to add to those in each cell starting at the perfectly proportional ($N = 1,172$) case.

Specifically, the problem is to find the number of observations for each state i and residence status j, X_{ij}, which can be added to the number of observations S_{ij} dictated by perfect proportionality (sample size of 1,172), given any chosen sample size, $N^* + T$, greater than 1,172 so as to:

$$\text{Minimize} \sum_{ij}\left(\frac{S_{ij} + X_{ij}}{N^* + T} - \alpha_{ij}\right)^2$$

$$\text{subject to} \sum_i \sum_j X_{ij} = T$$

$$S_{ij} + X_{ij} \leq Q_{ij}$$

$$X_{ij} \geq 0 \qquad \text{For all} \begin{cases} i = 1,\ldots,49 \\ j = 1,2 \end{cases}$$

where:

S_{ij} = $N^*\left(\alpha_{ij}\right)$ or 1,172 α_{ij}, the number of observations in each state/residency status cell which yields perfect proportionality

X_{ij} = The number of observations added to each state/residency status cell to augment the total sample size

T = A chosen sample size increment, $N^* + T \leq 11{,}168$

Q_{ij} = The number of observations in each state/residency status cell in the original unproportional mail survey sample reported in table 4-6.

α_{ij} = The "target" decimal frequency of occurrence of each state/residency status cell based on proportionality from table 4-5.

In this problem, everything is known except the X_{ij}'s. Its solution yields, in a sense, the best of a bad situation, since any allocation other than the proportional one will involve some divergence from the targeted frequencies, α_{ij}.

As a practical matter, we chose to modify the problem in order to solve it with the standard linear programming technique. The modification involves approximating the continuous nonlinear objective function for each ij with a discrete set of linear segments. Expanding the quadratic objective function:

$$OBJ_{ij} = \frac{S_{ij}^2 + 2S_{ij}X_{ij} + X_{ij}^2}{\left(N^* + T\right)^2} - 2\alpha_{ij}\frac{\left(S_{ij} + X_{ij}\right)}{N^* + T} + \alpha_{ij}^2$$

Eliminating constants and rearranging:

$$OBJ_{ij} = \frac{X_{ij}^2}{\left(N^* + T\right)^2} + \frac{2S_{ij}X_{ij}}{\left(N^* + T\right)^2} - \frac{2\alpha_{ij}X_{ij}}{N^* + T}$$

Differentiating and combining terms yields the expression:

$$\frac{\partial OBJ_{ij}}{\partial X_{ij}} = \frac{2\left(X_{ij} + S_{ij} - \alpha_{ij}\left(N^* + T\right)\right)}{\left(N^* + T\right)^2} \equiv c_{ij}\left(X_{ij}^k\right) \equiv C_{ij}^k$$

Evaluating of the above linear expression at K steps for each X_{ij}, and bounding each step by the interval ΔX_{ij}, produces a linear approximation to the continuous objective function.

The problem may then be transformed into the following linear programming problem:

$$\text{Min} \sum_i \sum_j \sum_k X_{ij}{}^k C_{ij}{}^k$$

$$\text{subject to } \Delta X_{ij}{}^k \geq X_{ij}{}^k \geq 0$$

$$\sum_i \sum_j \sum_k X_{ij}{}^k = T$$

and the intervals might be chosen as, for example:

$$\Delta X_{ij} = \frac{Q_{ij} - S_{ij}}{K}$$

and

$$\Delta X_{ij}{}^k = k \left[\frac{Q_{ij} - S_{ij}}{K}\right]$$

Results and Further Comment

The results of running the model described above are shown in figure B-1, where we have graphed the truncated objective function values for a range of subsample sizes from 4,000 to 8,000. The "optimum" is at a sample size of about 5,000, and that is the basis for the medium-sized, quasi-proportional sample used and referred to in the text. To judge from our own initial reaction to these results, however, it will be worthwhile to say something more about their interpretation and to give a brief heuristic explanation of how the process represented by the LP actually works.

First, it is potentially confusing to see that the objective function values are all negative and that therefore the minimum (at $N^* + T = 5{,}000$) is also farthest from the horizontal axis--i.e., is largest negatively. This is simply the result of our simplification and reflects the elimination of constants that, if reinserted, would shift everything up into the positive quadrant.

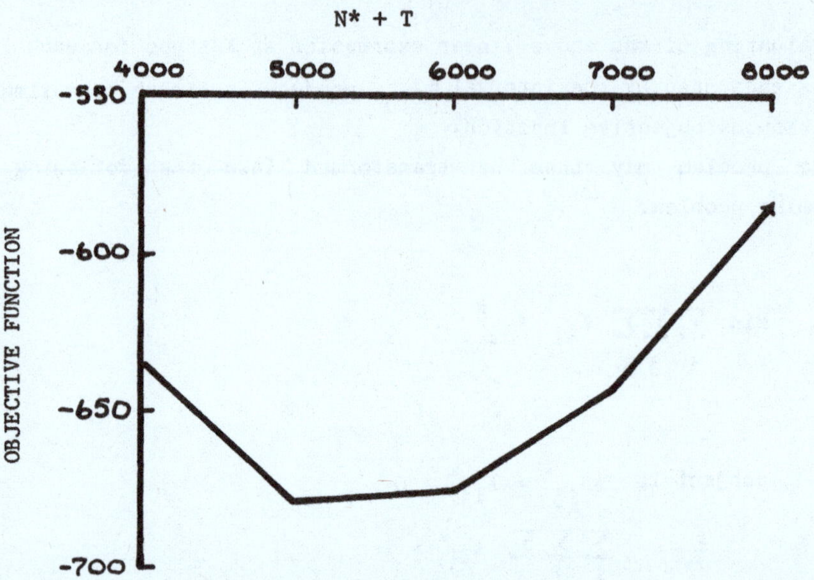

Figure B-1. Objective function values for various sample sizes.

Second, it may seem intuitively "clear" that the larger N* + T (hence the farther the subsample in total size from the proportional sample size), the worse the objective function result ought to be. We certainly asked ourselves how it is that the approach described above could produce a dip, with higher (worse) objective function values both near to and farther away from the proportional sample size. The short answer to this puzzle is that it is not only total subsample size that matters; it is, more importantly, how any particular subsample size can be spread around over the cells, given the cell-by-cell availabilities. As we increase the total subsample size we are looking at, it is true that in more and more cells we will go beyond the availability of data points. That is, for more and more cells it will be impossible to make

$$\frac{S_{ij} + X_{ij}}{N^* + T} = \alpha_{ij}$$

because $S_{ij} + X_{ij}$ will have exhausted the available observations, Q_{ij}. For these cells, as N* + T grows, the cell sample becomes more and more unproportional. On the other hand, for cells with many observations, over some ranges it is possible to bring the above proportion <u>closer</u> to α_{ij} as T grows and X_{ij} is judiciously increased.

Thus, even though it is true that no T can result in an objective function value better than that corresponding to the proportional sample (i.e., zero), it is also true that the objective function will not in general grow monotonically with N* + T. After an initial increase (departure from proportionality), the effect of additions to the X_{ij} in some cells can outweigh the effects of additions to N* + T in the denominator for all cells. Eventually, so many cells will have their available samples exhausted that the denominator effect will dominate, and the objective function value will rise (worsen) to the limit of N* + T.

A simple 4-cell example to help make these points is presented schematically in figure B-2. In this example, the sample and required proportions are such that the maximum proportional sample size is 10; though there are 20 available observations. The figure shows what happens to the objective function of the original nonlinear problem as total sample sizes above 10 are used, with the additional observations being added in such a way as to minimize the added "nonproportionality." The figure also shows how the additional observations are used and what happens to the actual α_{ij}'s in the process. The reader will see that for cell 1,1 the actual α_{ij} departs more and more from the desired proportion, 0,1. But for cells 2,1, 1,2 and 2,2, the actual proportion varies around the desired α_{ij} as the additions are made.

Having, we hope, disposed of the intuitive doubts raised by the results in figure B-1, we show in table B-1 how the "optimal" (best N* + T when T $\geq$ 2000 is required) sample size of 5,000 is distributed across the state and metro/nonmetro cells. This distribution serves as a guide for the actual drawing of the sample from the available observations, and the fractional cell sizes are translated by division by available cell observations into a fraction $0 \leq r_{ij} \leq 1.00$ that forms the cutoff for choosing or rejecting a particular observation to which a random number between zero and one has been attached. When at the 5,000 subsample size, the available observations for a particular cell are to be exhausted, r_{ij} = 1 and all observations are used.

Figure B-2. A simple example of finding "almost" proportional samples as total sample size varies

Available Sample N* + T = 20
Available Sample by cell

	1	2
1	1	6
2	3	10

Desired Cell Frequencies

	1	2
1	.1	.3
2	.2	.4

Implied Proportional Sample

N* = 10

	1	2
1	1	3
2	2	4

Steps in Proportioning N* + T

N* + T	OPTIMAL ADDITIONS			ACTUAL α_{ij}		OBJECTIVE FUNCTION VALUE
11	1	3		.091	.273	0.00413
	2	5	← (1)	.182	.455	
12	1	4	← (1)	.083	.333	0.00278
	2	5		.167	.417	
13	1	4		.077	.308	Minimum 0.00178
	(1) → 3	5		.231	.385	
14	1	4		.071	.286	0.00204
	3	6	← (1)	.214	.429	
15	1	5	← (1)	.067	.333	0.00222
	3	6		.200	.400	

Table B-1. Optimal Distribution of Observations for a 5,000-Observation Sample

State	Metropolitan fishermen Available	Metropolitan fishermen Optimal	Nonmetropolitan fishermen Available	Nonmetropolitan fishermen Optimal
ALABAMA	62	62.0*	116	63.5
ARIZONA	122	40.8	96	30.9
ARKANSAS	27	27.0*	172	60.5
CALIFORNIA	155	155.0*	44	44.0*
COLORADO	139	63.4	149	17.6
CONNECTICUT	89	57.8	123	14.1
DELAWARE	105	22.6	114	11.4
DISTRICT OF COLUMBIA	4	4.0*	0	0
FLORIDA	123	123.0*	80	59.3
GEORGIA	65	65.0*	117	87.0
IDAHO	55	23.2	236	28.1
ILLINOIS	199	166.8	254	62.8
INDIANA	80	80.0*	142	64.0
IOWA	72	46.0	139	63.4
KANSAS	56	42.4	164	56.2
KENTUCKY	36	36.0*	150	67.2
LOUISIANA	95	87.2	81	52.8
MAINE	33	20.2	204	24.9
MARYLAND	102	75.9	63	27.6
MASSACHUSETTS	84	84.0*	116	27.2
MICHIGAN	124	124.0*	118	76.4
MINNESOTA	96	96.0*	148	52.1
MISSISSIPPI	31	31.0*	117	64.5
MISSOURI	50	50.0*	143	65.0
MONTANA	82	17.2	221	25.7
NEBRASKA	62	33.0	165	37.0
NEVADA	104	23.2	143	15.2
NEW HAMPSHIRE	8	8.0*	230	24.8
NEW JERSEY	134	100.4	75	24.6
NEW MEXICO	49	20.8	181	20.8
NEW YORK	112	112.0*	85	70.6
NORTH CAROLINA	45	45.0*	133	86.2
NORTH DAKOTA	33	20.2	197	22.4
OHIO	106	106.0*	79	65.6
OKLAHOMA	106	59.0	88	57.2
OREGON	150	51.3	150	34.8
PENNSYLVANIA	111	111.0*	123	55.2
RHODE ISLAND	114	25.2	78	15.6
SOUTH CAROLINA	84	45.5	106	47.2
SOUTH DAKOTA	18	18.0*	201	23.7
TENNESSEE	39	39.0*	123	54.6
TEXAS	67	67.0*	121	90.4
UTAH	183	43.0	100	21.6
VERMONT	0	0	225	24.3
VIRGINIA	81	81.0*	112	54.5
WASHINGTON	251	88.6	415	50.5
WEST VIRGINIA	44	31.7	161	36.2
WISCONSIN	127	94.3	201	70.8
WYOMING	0	0	255	27.3
Total	4114	2823.7	7054	2177.3

Note: An asterisk, *, signifies that all available observations are used.

Appendix C

USING CASE WEIGHTS IN SAS REGRESSIONS

The case weight approach to the problem of making use of a non-proportional survey sample may be illustrated with a simple example. Suppose, for example, one has a survey sample with observations in each of four cells:

Table C-1

	State 1	State 2
Metro	n = 10	n = 20
Nonmetro	n = 30	n = 40

$\sum_n = 100 = N$

where n = number of responses in the survey from each of four cells

Suppose, also, that the population from which the survey was drawn is arrayed very differently, that is, the sample was designed in a nonrandom manner.

Table C-2

	State 1	State 2
Metro	n = 5,000	n = 50,000
Nonmetro	n = 2,000	n = 10,000

$\sum_n = 67,000 = N$

Obviously, the relative proportions in the four cells are very different. If $f_i = n_i/N$, then the population and sample proportions are as follows:

Table C-3

	Population			Sample	
	State 1	State 2		State 1	State 2
Metro	0.075	.75	Metro	0.10	0.20
Nonmetro	0.03	0.145	Nonmetro	0.30	0.40

To correct this situation, one can derive a set of caseweights to proportionalize the sample (note that this has no direct relationship to weights used for heteroskedasticity or autocorrelation corrections, although methods used to implement a caseweight correction are superficially similar to "error-term" correction methods.) To compute the caseweights, let f_{pop} be the relative frequency of occurrence in a sample cell. In the example, f_{pop} (metro, state 1) = 0.075, and f_{samp} (metro, state 1) = 0.10. The appropriate weight, D (metro, state 1) =

$$D_{m',1} = \frac{f_{pop}\left(\text{Metro, State 1}\right)}{f_{samp}\left(\text{Metro, State 1}\right)} = \frac{0.075}{0.10} = 0.75$$

Tabulated D values for the example are:

Table C-4

	State 1	State 2
Metro	0.750	3.750
Nonmetro	0.100	0.362

Inspection of the tables indicates that, for example, there are "too many" state 1 nonmetro observations in the sample, and "too few" state 2 metro observations.

The next question is how actually to make the sample representative, using the numbers in table C-4 in a SAS OLS regression. Three methods suggest themselves. First, we can replicate observations in each cell so that the sample is proportional. To do this, one would first multiply each of the four numbers in table C-4 by 1,000. This would make each entry an integer as in table C-5.

Table C-5

	State 1	State 2
Metro	750	3750
Nonmetro	100	362

Thus, in the new, literally replicated sample, each observation for state 1 metro would be repeated 749 (750-1) times, state 1 nonmetro 99 times, etc. The survey sample would then be proportional to the population from which it was drawn. However, two problems make this method both impractical and misleading for statistical significance testing of the regression and its estimated parameters. First, given the magnitude of the numbers of table C-5, one ends up with an enormous number of observations ($\approx$32,000 in this example). More important, this method overstates the precision of both the estimated β's and the overall regression. The degrees of freedom in the regression will be incorrect, as will the standard errors of the estimated parameters. The problem arises because one has multiplied the number of observations with no increase in the amount of information in the sample. Therefore, this method should not be used as a cure for sample balance problems.

The second method involves transformation of each variable in each observation by $\sqrt{D}$ corresponding to that observation. In the example, these values are:

Table C-6

	State 1	State 2
Metro	0.87	1.94
Nonmetro	0.32	0.60

Thus each variable in the state 1, metro cell would be transformed (multiplied) by 0.87; similarly for the other cells. This method produces correct β's and standard errors of the coefficients in the regression, but incorrrect coefficients of determination $\left(R^2\text{'s}\right)$ (Blomquist, 1980). Obviously, if R^2 interpretation is important in an investigation, this method should not be employed.

Finally, one can "proportionalize" the sample by using the SAS WEIGHT statement. This can be done by using "D" as a weight in a weighted least squares regression. This method allows one to compute a regression equation with <u>correct</u> β's, standard errors, and R^2's, and is the preferred method whenever it is possible.

References

N. Soren Blomquist. 1980. "A Note on the Use of Coefficient of Determination," Scandinavian Journal of Economics vol. 82, no. 3, pp. 409-412.

Appendix D

METHODS FOR DETECTING AND CORRECTING FOR HETEROSKEDASTICITY

Theoretical considerations indicate that heteroskedasticity will be present in equations estimated from the fee-fishery survey data. This appendix outlines the method for detecting the exact nature of the heteroskedasticity and subsequently correcting for it. The emphasis here is on practical solutions to an existing problem; see chapter 3 of Goldfeld and Quandt (1972) for a concise discussion of the theory involved.

We use the Park method (Park, 1966) for detection of heteroskedasticity in the fee-fishing data set. Park posits the following structure for the variance of the error term in a heteroskedastic OLS regression:

$$\sigma_i^2 = \sigma^2 Z_i^{\gamma} e^{V_i}$$

so that

$$\ell n\sigma_i^2 = \ell n\sigma^2 + \gamma \ell n Z_i + V_i$$

where V_i has "nice" properties (mean 0, zero covariance, and so forth). In our case, this means that one does a regression of the following form:

$$\ell n\hat{e}_i^2 = \hat{a} + \hat{b}_i \left(\ell n X_i\right) + V_i$$

$\hat{e}_i^2$ = squared residual from the initial OLS regression

X_i = an independent variable from the initial regression.

Because the Park method allows one both to test and correct for heteroskedasticity if detected, we employ it in our travel cost regressions. The Park test results produced from our fee-fishing data are presented in table D-1. They are obtained by regressing the log of the squared residuals (LNRSQ) generated from a particular visitation-travel cost OLS regression on the logs of travel cost, per capita income and population (LNTC, LNCAPINC, LNPOP, respectively). Based on the F test of the coefficient of determination, we can say that for all models there is a significant relationship between the error term and the explanatory variables, taken as a group. As the explanatory variables are potentially collinear, the weights matrices used to estimate all pseudo-GLS models discussed in the "Results" section of the text were generated using all three estimated slope parameters to generate weights.

References

Goldfeld, Stephen M. and Richard E. Quandt. 1972. Nonlinear Methods in Econometrics (New York, North-Holland).

Kmenta, Jan. 1971. Elements of Econometrics (New York, MacMillan).

Park, R. E. 1966. "Estimation with Heteroskedastic Error Terms," Econometrica vol. 34, no. 4, p. 888.

Table D-1. Park Test Results for Trout and Catfish OLS Regressions

MODEL: TROUT.1 SSE : 2498.326 F RATIO : 7.04
DEP VAR: LNRSQ DFE : 467 PROB > F : 0.0002
MSE : 5.349735 R-SQUARE : 0.0432

Variable	DF	Parameter Estimate	Standard Error	T Ratio	PROB > 1 T 1
INTERCEPT	1	-13.175	6.985	-1.886	0.0599
LNTC1	1	-0.377	0.249	-1.513	0.1309
LNCAPINC	1	2.863	0.854	3.469	0.0006
LNPOP	1	-0.0218	0.077	-0.283	0.7771

MODEL: TROUT.2 SSE : 2588.416 F RATIO : 6.53
DEP VAR: LNRSQ DFE : 467 PROB > F : 0.0003
MSE : 5.42647 R-SQUARE : 0.0402

Variable	DF	Parameter Estimate	Standard Error	T Ratio	PROB > 1 T 1
INTERCEPT	1	-6.760	7.080	-0.955	0.340
LNTC2	1	-0.691	0.240	-2.876	0.004
LNCAPINC	1	2.270	0.868	2.614	0.009
LNPOP	1	0.048	0.090	0.536	0.592

MODEL: CAT.1 SSE : 1918.885 F RATIO : 6.02
DEP VAR: LNRSQ DFE : 312 PROB > F : 0.0006
MSE : 6.15027 R-SQUARE : 0.0547

Variable	DF	Parameter Estimate	Standard Error	T Ratio	PROB > 1 T 1
INTERCEPT	1	12.282	7.663	1.608	0.1100
LNTC1	1	-1.059	0.252	-4.212	0.0001
LNCAPINC	1	0.044	0.938	0.047	0.9623
LNPOP	1	0.145	0.091	1.597	0.1113

MODEL: CAT.2 SSE : 2084.85 F RATIO : 5.49
DEP VAR: LNRSQ DFE : 312 PROB > F : 0.0012
MSE : 6.682212 R-SQUARE : 0.0502

Variable	DF	Parameter Estimate	Standard Error	T Ratio	PROB > 1 T 1
INTERCEPT	1	11.756	7.997	1.470	0.1426
LNTC1	1	-1.077	0.268	-4.012	0.0001
LNCAPINC	1	0.092	0.979	0.094	0.9248
LNPOP	1	0.217	0.109	1.993	0.0471